CAMBRIDGE TRACTS IN MATHEMATICS

---

72. *Completeness and basis properties of sets of special functions*

J. R. HIGGINS

Lecturer in Mathematics
The Cambridgeshire College of Arts and Technology

72. Completeness and basis properties of sets of special functions

D0169055

CAMBRIDGE UNIVERSITY PRESS

CAMBRIDGE

LONDON · NEW YORK · MELBOURNE

WILLIAM MADISON RANDALL LIBRARY UNC AT WILMINGTON

Published by the Syndics of the Cambridge University Press
The Pitt Building, Trumpington Street, Cambridge CB2 1RP
Bentley House, 200 Euston Road, London NW1 2DB
32 East 57th Street, New York, NY 10022, USA
296 Beaconsfield Parade, Middle Park, Melbourne 3206, Australia

First published 1977

Printed in Great Britain at the
University Press, Cambridge

Library of Congress Cataloguing in Publication Data
Higgins, John Rowland, 1935–
Completeness and basis properties of sets of special functions.
(Cambridge tracts in mathematics; 72)
Bibliography: p.
Includes indexes.
1. Functions, Special.   2. Series, Orthogonal.
3. Lp spaces.   I. Title.   II. Series.
QA351.H57   515'.5   76-19630
ISBN 0 521 21376 2

To Nan and P.B.

Contents

[vii]

Preface

Some years ago I came across the need for precise information concerning the basis properties of sets of special functions, and the methods available for testing for such properties. This material proved to be rather widely scattered, so I began a collection of notes on the subject which have formed the foundations of the present little book.

I hope that the book will prove useful to graduate students of mathematics, particularly those whose research interests are developing in the direction of bases in Hilbert and Banach spaces: it could bridge the gap that exists between the scant treatment this topic usually receives in standard texts on functional analysis on the one hand, and the rather formidable specialist books such as Marti (1969) and Singer (1970) on the other. There is no harm in having some experience on the practical side of the business before aiming to become managing director!

I hope the book will appeal to workers in other scientific fields as well. An appendix has been included which lists many of the standard results, and this may help to make the book useful as a source of reference.

It is assumed that the reader's education will have included the usual first courses in real variable (including Lebesgue integration) and complex variable. Although a knowledge of functional analysis would be an advantage it is not strictly necessary, and all the principles of functional analysis which are used in the text are listed in an appendix, along with certain other facts which do not fit easily into the presentation.

Introductory sections on metric and L^p spaces have been included, in note form since it is assumed that most readers will already be familiar with this material. These sections are intended to serve as a 'run up' to the main part of the book.

The subject matter touches upon many important topics in both pure and applied mathematics; for example, bases in Banach space, orthogonal series, properties of special functions, interpolation and approximation, eigenfunctions and boundary value problems, probability, and information theory. These, together with a variety of methods of proof both ancient and modern, give the subject a certain charm. This is a source of satisfaction to me, and I hope it will prove equally satisfactory to the reader; if so, my work will have been well rewarded.

I wish to express my gratitude for having been allowed some remission of teaching duties at the Cambridgeshire College of Arts and Technology for purposes of writing and research over a three-year period, during which parts of the book were written. I am particularly grateful to those of my colleagues who, as a consequence, had to shoulder an extra burden of work.

I would also like to thank Dr F. Smithies, fellow of St John's College, Cambridge, for reading the manuscript and making many valuable suggestions; as a result of this the book has been improved in every respect. Appreciation is also due to the staff of Cambridge University Press for accepting the book as a 'tract', and for their courteous efficiency.

Cambridge   J. R. HIGGINS
March 1976

1. Foundations

The sets of functions which form the subject matter of this book are to be considered as sequences in *metric spaces*. Actually we shall be almost exclusively concerned with various L^p spaces, particularly the case $p = 2$, and with subspaces of such spaces. Although the notes which follow in § 1.1 and § 1.2 contain sufficient metric space theory for an understanding of the rest of the book, the reader who is new to metric spaces may wish to fill in from a good text such as Copson (1967). For general background reading Simmons (1963) is also highly recommended.

Throughout the book an effort has been made to present theorems which are sufficiently general to be 'useful', but in a small introductory book of this kind a great deal of detail has to be left out; adequate references are given for those who want to consult more advanced sources.

1.1 Notes on metric spaces

1.1.1 Vector space It is assumed that the reader is familiar with elementary set theory. The word 'space' is used in mathematics to mean a set with some internal 'structure'. Let V be a set whose elements are to be called *vectors*, and let F be a field (the *field of scalars*; we will usually take it to be the field of complex numbers). The basic structure that we shall require for V is that it be closed under an operation of addition of two vectors u and v, denoted by $u+v$, and an operation of multiplication of a vector u by a scalar f of F, denoted by fu. If V is to become a useful mathematical system we shall require more structure within it than the presence of these two operations. If in addition the two operations satisfy the following list of axioms, then V is called a *vector space*.

(1) $u+v \in V$ for every u and v in V (V is closed under the operation of addition).

(2) $u+(v+w) = (u+v)+w$ for every u, v, w in V (addition is associative).

(3) $u+v = v+u$ for every u and v in V (addition is commutative).

(4) V contains a vector θ such that $u+\theta = u$ for every u in V (θ is the 'null vector' of V).

(5) For every $u \in V$ there is a vector $-u$ in V such that $u+(-u) \equiv u-u = \theta$ (each vector has an additive inverse).

(6) fu is in V for every f in F and every u in V (V is closed under multiplication by scalars).

(7) Let 1 denote the multiplicative unit of F. Then $1u = u$ for every u in V.

(8) For every u and v in V and f and g in F we have

$$f(u+v) = fu+fv$$

and $\qquad\qquad (f+g)u = fu+gu$ (distributive laws).

(9) $(fg)u = f(gu)$ for every f and g in F and u in V.

The first five axioms express the Abelian group character of V.

From now on, V will denote a vector space; the definitions to follow give further structure to V.

1.1.2 Metric and norm A *metric* on V is a real valued 'distance' function ρ, defined on pairs (u, v) of vectors in V, such that for every u, v, w in V we have

(1) $\rho(u, v) \geqslant 0$, and $\rho(u, v) = 0$ if and only if $u = v$;

(2) $\rho(u, v) + \rho(v, w) \geqslant \rho(u, w)$ (triangle inequality);

(3) $\rho(u, v) = \rho(v, u)$.

Note that a space need not be a vector space in order to define a metric on it.

A *norm* on V is a real valued function defined on V and denoted by $\| \ \|$, such that

(1) $\|u\| \geqslant 0$, $\|u\| = 0$ if and only if $u = \theta$;

(2) $\|u+v\| \leqslant \|u\| + \|v\|$ (triangle inequality);

(3) $\|fu\| = |f| \, \|u\|$ for every complex number f.

The norm generalises the notion of absolute value of complex numbers. Evidently the choice $\rho(u,v) = \|u-v\|$ provides a metric on V, and this special metric is called the *metric induced by the norm*. A space possessing a metric is called a *metric space* and one possessing a norm is called a *normed space*. From now on unless otherwise stated V will denote a normed vector space with metric induced by the norm.

1.1.3 Convergence Let (u_n) be a sequence (see §1.1.10) of elements of V. Then (u_n) is said to be *convergent* if there exists u in V, called the *limit* of the sequence, such that $\|u_n - u\| \to 0$ as $n \to \infty$. A convergent sequence has a unique limit. The reader must notice carefully that convergence means convergence in the norm of V to an element of V. For example, one can construct a sequence of rationals which 'converges' to $\sqrt{2}$, but here convergence must be understood to take place within the normed vector space of real numbers, even though each member of the sequence is a rational number. If one were to speak only of the rationals, the sequence would not be convergent; it would be a ' Cauchy sequence', however, and these ideas lead to a most important and desirable property of metric spaces, that of 'completeness' (see § 1.1.5).

1.1.4 Closed sets Let $S \subset V$ and $u \in V$. Then u is called a *point of closure* of S if, given $\epsilon > 0$, there exists an $s \in S$ such that $\|u - s\| < \epsilon$. The *closure* $\bar{S}$ of S is the collection of all points of closure of S. S is called *closed* if $S = \bar{S}$. For every S we have $S \subset \bar{S} = \bar{\bar{S}}$.

Let (u_n) be a sequence in V. Any expression formed from vectors of this sequence by use of the two basic operations of addition and of multiplication by a scalar is called a *linear combination* of those vectors. The collection U of all such finite linear combinations is called the *linear span* of (u_n) and $\bar{U}$ is called the *closed linear span* of (u_n) frequently denoted by $[u_n]$. One sometimes says that (u_n) *spans* U.

1.1.5 Cauchy sequences and completeness The sequence (u_n) in V is said to be a *Cauchy sequence* if, given $\epsilon > 0$, there exists N such that $\|u_n - u_m\| < \epsilon$ for every n and m greater than N. V is said to be *complete* if every Cauchy sequence in it converges.

1.1.6 Dense subsets A subset $S \subset V$ is said to be *dense* in V if for every $u \in V$ and $\epsilon > 0$, there exists a vector $s \in S$ such that $\|u - s\| < \epsilon$. The reader may verify that if $S_1 \subset S_2 \subset V$ with S_1 dense in S_2 and S_2 dense in V, then S_1 is dense in V. We shall refer to this as the *chain of dense subsets* principle. If V contains a countable dense subset then it is called *separable*.

1.1.7 Banach space In his well-known book, *Théorie des opérations linéaires*, Stefan Banach (1932) referred to certain spaces as '*les espaces du type (B)*' and such spaces have carried his name ever since; we are now in a position to write down the definition: a normed vector space which is complete in the metric induced by the norm is called a *Banach space*.

1.1.8 Hilbert space Let V be a vector space. A complex valued function defined on pairs of vectors of V is called an *inner product*, and written (u, v) for u and v in V, if it satisfies

(1) $(u_1 + u_2, v) = (u_1, v) + (u_2, v)$ $(u_1, u_2, v \in V)$;

(2) $(cu, v) = c(u, v)$ for every $c \in F$; ((1) and (2) express 'linearity' in the first argument);

(3) $(u, v) = \overline{(v, u)}$ (here a bar denotes complex conjugate; this is the 'Hermitian' symmetry property of the inner product);

(4) $(u, u) \geqslant 0$, and $(u, u) = 0$ if and only if $u = \theta$.

A space V in which each pair of vectors has an inner product is called an *inner product space*. Note that by (3) (u, u) is real; in order to construct a norm for V we may put $(u, u)^{\frac{1}{2}} = \|u\|$, and this choice does indeed provide a norm (see problem 1.4).

THEOREM (Schwarz' inequality) *Let V be an inner product*

space with norm given by $\|u\| = (u, u)^{\frac{1}{2}}$. *Then* $|(u, v)| \leqslant \|u\| \, \|v\|$
for every u and v in V.

Proof Assume that neither u nor v is null, since in this case
the theorem is obvious. For any scalar c we have $\|u + cv\|^2 \geqslant 0$.
As the proof proceeds we shall see how to choose an appropriate
c. We have

$$0 \leqslant (u + cv, u + cv)$$
$$= \|u\|^2 + |c|^2 \|v\|^2 + (cv, u) + (u, cv)$$
$$= \|u\|^2 + |c|^2 \|v\|^2 + 2 \operatorname{Re} c(u, v) \text{ by (1), (2) and (3) above.}$$

Choose $\arg c$ such that $c(u, v)$ is real and negative; i.e. choose
$\arg c = \pi - \arg (u, v)$. Then

$$\|u\|^2 + |c|^2 \|v\|^2 \geqslant - 2c(u, v) = 2|c| \, |(u, v)|.$$

Now choose $|c| = \|u\|/\|v\|$, which yields the required result.

Let V be an inner product space. If V is a Banach space with
respect to the norm defined by $\|u\| = (u, u)^{\frac{1}{2}}$, then V is said to
be a *Hilbert space*. Thus the Hilbert spaces form a subclass of the
Banach spaces.

The reader cannot fail to have noticed that metric spaces
appear to have 'geometrical' properties analogous to properties
of ordinary finite dimensional Euclidean space. For example, the
metric itself provides the notion of distance in V, the norm
gives the distance from the 'origin' and is thus some measure-
ment of the size of an element, and in Hilbert space the inner
product generalises the dot product of ordinary vectors. Indeed,
the bases which are the subject of this book are nothing but
generalisations of the bases of unit vectors of finite dimensional
vector spaces. The reader will find many more such geometrical
facts in the following pages, and is enjoined to develop a geometrical way of thinking about Hilbert and Banach space.

Problems

1.1 Show that in a metric space the triangle inequality
$$\|u + v\| \leqslant \|u\| + \|v\|$$
is equivalent to $\quad \|u - v\| \geqslant |\|u\| - \|v\||.$

1.2 Show that one has equality in Schwarz' inequality if and only if u and v are linearly dependent.

1.3 Formulate and prove a Pythagoras theorem in Hilbert space.

1.4 Show that the choice $\|u\| = (u,u)^{\frac{1}{2}}$ provides a norm for the inner product space V.

1.5 Prove the 'parallelogram identity'

$$\|u+v\|^2 + \|u-v\|^2 = 2\|u\|^2 + 2\|v\|^2$$

in Hilbert space, and justify the name 'parallelogram'.

1.6 Show that if the norm of a Banach space satisfies the parallelogram identity then it is a Hilbert space. Hint: introduce an inner product by use of the 'polarisation identity'

$$(u,v) = \tfrac{1}{4}(\|u+v\|^2 - \|u-v\|^2 + i\|u+iv\|^2 - i\|u-iv\|^2).$$

1.7 Show that the norm is a continuous function on a Banach space to $\mathbb{R}$.

1.1.9 The projection theorem Let H be a Hilbert space. A *linear manifold* or *subspace* of H is a subset which is algebraically closed under the operation of taking linear combinations of its elements. The reader should check that this definition does lead to what he would expect from the word 'subspace'. A linear manifold which is closed as a subset of H is called a *closed linear manifold*.

Two vectors of H are said to be *orthogonal* if their inner product is zero; a vector is said to be orthogonal to a subspace if it is orthogonal to every vector of that subspace. The *orthogonal complement* $S^{\perp}$ of a subset S of H is the collection of all vectors orthogonal to S.

Let S and T be subspaces of H such that $S \cap T = \{\theta\}$; then the set of all vectors of the form $u + v$ with u in S and v in T is called the *direct sum* of S and T and is written $S \oplus T$. The following theorem is a most satisfying result, and appeals to our geometrical way of thinking about Hilbert space.

THEOREM (Projection theorem) *Let S be a subspace of Hilbert space H. Then $S \oplus S^{\perp} = H$.*

For the proof, see Yosida (1965) p. 82.

1.1.10 Mappings The words mapping, function, functional, operator, transformation, etc. occur repeatedly in mathematical writing and, whilst their meanings have become fairly standard, no completely satisfactory standardisation has so far evolved. We shall adopt the following definitions.

A *function* on, *mapping* or *transformation* of, a set V into a set W is a rule f which assigns a unique $w \in W$ to each $v \in V$. This association will be denoted by the usual functional notation $w = f(v)$. V is called the *domain* of f and $W_1 = \{f(v) : v \in V\}$ is called the *range* of f. The notation $f \colon V \to W$ is used to mean 'f maps V into W'. Sometimes we call w the *image of v by f*, and this notation carries over to sets; thus the set

$$\{f(v) : v \in A \subset V\} = f(A)$$

is called the *image of A by f*.

If $W_1 = W$ then f is called a mapping *onto* W.

A mapping f is called *one-to-one* if $v_1 \neq v_2$ implies $f(v_1) \neq f(v_2)$.

If V is a vector space, a mapping f of V is called *linear* if it preserves the two basic operations of addition and multiplication by scalars, that is, if $f(c_1 v_1 + c_2 v_2) = c_1 f(v_1) + c_2 f(v_2)$ for every c_1, c_2 in F and v_1, v_2 in V.

A *functional* is a mapping of V to $\mathbb{R}$. If V is a normed vector space a functional f on V is called *bounded* if there exists a real number c such that, for every $v \in V$, $|f(v)| \leqslant c\|v\|$; the infimum of all such cs is called the *norm* of f and written $\|f\|$.

If $f \colon V \to V$ then f is called an *operator* on V.

A *sequence* in W is a mapping of a subset J of $\mathbb{R}$ into W. Usually, but not always (see, e.g. § 2.2), J, the *indexing set*, is such that $J \subseteq \mathbb{N}$. If a sequence maps $n \in J$ to $\phi_n \in W$ then it is denoted by $(\phi_n)_{n \in J}$.

A mapping f between two normed spaces is called an *isometry* if it is one-to-one and norm preserving, that is, whenever $w = f(v)$ then $\|w\| = \|v\|$.

A mapping which is one-to-one, linear and onto is called an *isomorphism*. A mapping which is both isometric and isomorphic is called an *isometric isomorphism*. An operator which is also an isometry on a Hilbert space is called a *unitary operator*.

1.1.11   The dual of a Banach space, strong and weak convergence   Let B be a Banach space. The class B^* of all bounded linear functionals on B is also a Banach space whose norm is the linear functional norm defined in § 1.1.10. B^* is called the *dual space* of B. One can define a linear functional on the dual space by the process of fixing v in B and forming

$$F_v(f) = f(v)$$

where f varies over B^*. Then F_v is obviously a bounded linear functional on B^* for every $v \in B$. It may turn out that B^* admits no other bounded linear functionals, in which case there is a natural one-to-one association of points $v \in B$ with points F_v in $(B^*)^* = B^{**}$, the *second dual* space of B. Banach spaces with this property are called *reflexive*. The L^p spaces (see § 1.2) are reflexive Banach spaces for $p > 1$, but not for $p = 1$.

Convergence in B is often called strong convergence, that is, the sequence (v_n) *converges strongly* to v if $\|v_n - v\| \to 0$ as $n \to \infty$. There is a companion mode of convergence in B called weak convergence, associated with the linear functionals on B. The sequence (v_n) *converges weakly* to v if $|f(v_n) - f(v)| \to 0$ as $n \to \infty$ for every $f \in B^*$. *Strong convergence implies weak convergence to the same limit*, for

$$|f(v_n) - f(v)| = |f(v_n - v)| \leqslant \|f\| \, \|v_n - v\|.$$

The converse is not true (find an example to show this!).

1.2   Notes on the L^p spaces

In this section we present some of the basic facts about the L^p spaces, again in note form.

Let X denote any measurable subset of $\mathbb{R}$, of finite or infinite Lebesgue measure. For any real number p such that $1 \leqslant p < \infty$, we are going to consider the class $L^p(X)$ of complex valued functions whose pth powers are Lebesgue measurable and integrable over X. These classes can be generalised in various ways, e.g. by defining the functions on a σ-finite measure space.

Let p be given, and f lie in the class described above. Put

$$\|f\|_p = \left\{\int_X |f|^p\right\}^{1/p}.$$

We have deliberately chosen the notation of the norm here. We shall also need $L^\infty(X) = \{f : f$ bounded except possibly on a set of measure zero, measurable on a set X of finite measure, and ess sup $|f| = \lim_{p \to \infty} \|f\|_p < \infty\}$.

HÖLDER'S INEQUALITY   *Let* $f \in L^p(X)$, $g \in L^q(X)$, $1/p + 1/q = 1, p \geqslant 1$. *Then* fg *is integrable over* X, *that is,* $fg \in L^1(X)$, *and*

$$\|fg\|_1 \leqslant \|f\|_p \|g\|_q.$$

Equality holds if and only if f and g are linearly dependent.

MINKOWSKI'S INEQUALITY   *Let both f and g belong to* $L^p(X), p \geqslant 1$. *Then* $f + g \in L^p(X)$, *and*

$$\|f + g\|_p \leqslant \|f\|_p + \|g\|_p$$

Equality holds if and only if f and g are linearly dependent.

There are various extensions of these inequalities, for example to the case of more than two functions, and to other values of p (when $p < 1$ the inequalities are reversed; see Hardy, Littlewood and Polya (1952)).

For a given p, the class of functions and the formula for $\|f\|_p$ described above yield a Banach space $L^p(X)$ and its norm, provided that one more stipulation is made. This is that two functions f and g whose values differ only on a set of measure zero must be considered as the same Banach space element, since otherwise we should have more than one null element and this would contradict the definition of the norm. Thus $L^p(X)$ consists of equivalence classes, two functions being in the same equivalence class if they differ only on a set of measure zero.

Minkowski's inequality is the triangle inequality for the norm.

When $p = 1$ we frequently omit the '1' from notation. When $1 < p < \infty$ $L^p(X)$ is a separable, reflexive Banach space whose dual space is isometrically isomorphic to $L^q(X)$, $1/p + 1/q = 1$; the completeness is a well-known theorem of Riesz and Fischer.

From Hölder's inequality we have $L^p(X) \subset L^r(X)$ if $r < p$ and X is of finite measure.

In the special case $p = 2$ we have

$$\|f\|_2^2 = \int_X |f|^2 = \int_X f\bar{f}.$$

We can define an inner product on $L^2(X)$ by putting

$$(f, g) = \int_X f\bar{g} < \infty;$$

then $L^2(X)$ becomes a Hilbert space. Note that Schwarz' inequality for this Hilbert space is a special case of Hölder's inequality. We shall also need the Hilbert space $L^2(X, w)$ of (equivalence classes of) functions which are square integrable with respect to the 'weight function' w, i.e. those f for which

$$\int_X |f|^2 w < \infty \text{ (see pp. 28–33)}.$$

The weak convergence of the sequence (v_n) to v in $L^2(X)$ is equivalent to the condition

$$(w, v_n) \to (w, v) \quad (w \in L^2(X)).$$

The continuous functions are dense in $L^p(a, b)$, $1 \leqslant p < \infty$, for a finite interval (a, b) (so are they in $L^p(\mathbb{R}^n)$). The class $C(a, b)$ of all continuous functions on a closed finite interval $[a, b]$ is a Banach space with norm $\|f\| = \sup\{|f(x)| : x \in [a, b]\}$. Then the Weierstrass approximation theorem (Appendix 1,3(a)) says that the polynomials are dense in C. Now if $f \in C$, then $f \in L^p(a, b)$ and a simple calculation shows that the polynomials are also dense in C in the L^p norm. From the chain of dense subsets principle (§ 1.1.6) it follows that the polynomials are dense in $L^p(a, b)$, $1 \leqslant p < \infty$. Note that $[a, b]$ must be a *finite* interval. We can now state an important result: *The set $\{\Sigma a_n x^n\}$ of all linear combinations of powers $\{x^n : n = 0, 1, 2, \ldots\}$ is dense in $L^p(a, b)$, $1 \leqslant p < \infty$.* We shall return to this result and generalise it in § 2.1.

1.3 Orthogonal sequences in Hilbert space

1.3.1 The theorems of Riesz–Fischer, Bessel and Parseval

From now on H will always denote a Hilbert space, with inner product $(.,.)$. Let J be an indexing set, and let $(\phi_n)_{n \in J}$ be a sequence with $\phi_n \in H$ $(n \in J)$. This indexing set may have the cardinality of the continuum but we shall see later, in the context of the 'dimension' of H (p. 18), that it is usually sufficient that J be a countable set (usually a set of integers).

DEFINITION The sequence (ϕ_n), $\phi_n \neq \theta$, in H is called *orthogonal* if $(\phi_n, \phi_m) = 0$ for $n \neq m$. It is called *normal* if every ϕ_n has unit norm. These two properties are conveniently expressed by using the Kronecker delta symbol, δ_{nm}, which equals 0 if $n \neq m$ and equals 1 if $n = m$; then a sequence which is orthogonal and normal satisfies $(\phi_n, \phi_m) = \delta_{nm}$. Such sequences are called *orthonormal* (written ON for short).

In this definition we have omitted mention of the indexing set and will continue to do so when this will not cause confusion. Also, it is not expected to cause confusion if we refer to the *set* $\{\phi_n\}$, meaning the range of the sequence (ϕ_n). Thus we speak of 'the expansion in the *set* $\{\phi_n\}$', or again of a '*set* of special functions forming an ON sequence ...', etc.

Let (ϕ_n) be any ON sequence in H, and $(\phi_n)_{1 \leqslant n \leqslant N}$ any finite subsequence relabelled with integer subscripts if necessary.

LEMMA (Best approximation) *For any* $f \in H$

(i) $\left\| f - \sum\limits_{n=1}^{N} a_n \phi_n \right\|$ *is smallest when* a_n *has the value* $c_n = (f, \phi_n)$ *for every* n.

(ii) $\sum\limits_{n=1}^{N} |c_n|^2 \leqslant \|f\|^2$.

Where no confusion can arise the running index will be omitted from summation signs.

To prove (i) we have

$$0 \leqslant \|f - \Sigma\, a_n\, \phi_n\|^2$$
$$= \|f\|^2 - 2\,\mathrm{Re}\,\Sigma\, \bar{a}_n (f, \phi_n) + \Sigma |a_n|^2$$
$$= \|f\|^2 - 2\,\mathrm{Re}\,\Sigma\, \bar{a}_n c_n + \Sigma |a_n|^2$$
$$= \|f\|^2 + \Sigma |a_n - c_n|^2 - \Sigma |c_n|^2.$$

This is obviously smallest for $a_n = c_n$, and with this choice (ii) follows.

The quantities $c_n = (f, \phi_n)$ are called the *Fourier coefficients* of f with respect to (ϕ_n); the terminology is obviously borrowed from the classical Fourier analysis in which (ϕ_n) is the well-known ON sequence of trigonometrical functions. We show next that *f can have at most countably many non-zero Fourier coefficients.* For if we are given any f in H and any positive integer k, we can choose the positive integer N so large that $\|f\|^2 < N/k$. Then by part (ii) of the previous lemma we have

$$\sum_{n=1}^{N} |(f, \phi_n)|^2 < N/k$$

for all possible ways of selecting N members of the sequence $(\phi_n)_{n \in J}$. Thus, the number of ϕ_ns for which $|(f, \phi_n)|^2 > 1/k$ is less than N. Since k is arbitrary the required result follows.

We now return to conclusion (i) of the previous lemma and assume that f lies in $[\phi_n]_{n \in J}$, i.e. given $\epsilon > 0$ there is some sequence of coefficients $(a_n)_{1 \leqslant n \leqslant N}$ such that

$$\left\| f - \sum_{n=1}^{N} a_n \phi_n \right\| < \epsilon.$$

Evidently $a_n = c_n$ is an admissible choice regardless of the ϵ chosen; since there are only countably many c_ns associated with f we may let $N \to \infty$ and obtain

$$\lim_{N \to \infty} \left\| f - \sum_{n=1}^{N} c_n \phi_n \right\| = 0.$$

This expresses the fact that f has been expanded in a *Fourier series* in the sequence (ϕ_n), convergence being in the strong sense.

We shall now show that f can have at most one such expansion, that is, that f *determines its Fourier coefficients uniquely by the formula* $c_n = (f, \phi_n)$.

For if (a_n) be any sequence of coefficients for which

$$\|f - \Sigma\, a_n\, \phi_n\| \to 0,$$

then for each n we have

$$|c_n - a_n| = |(f, \phi_n) - a_n|$$

$$= \left|\left(f - \sum_{m=1}^{N} a_m\, \phi_m,\, \phi_n\right)\right| \text{ for all } N \geqslant n,$$

$$\leqslant \left\|f - \sum_{m=1}^{N} a_m\, \phi_m\right\| \|\phi_n\|.$$

We may let $N \to \infty$ and conclude that $a_n = c_n$ for every n.

This does not rule out the existence of an element g, distinct from f, for which $c_n = (g, \phi_n)$; actually we shall see that this cannot happen if there are 'sufficiently many' ϕ_ns (see the Riesz–Fischer theorem, § 1.3.2).

We may now return to conclusion (ii) of the 'best approximation' lemma, and there too let $N \to \infty$. We obtain

BESSEL'S INEQUALITY   *Let* $f \in H$, $(\phi_n)_{n \in J}$ *be an ON sequence in* H *and* (c_n) *the (necessarily countable) sequence of Fourier coefficients of* f *with respect to* (ϕ_n). *Then*

$$\sum_{n=1}^{\infty} |c_n|^2 \leqslant \|f\|^2.$$

A glance at the proof of conclusion (i) of the best approximation lemma yields

PARSEVAL'S THEOREM   *With the notations of Bessel's inequality, we have*

$$\sum_{n=1}^{\infty} |c_n|^2 = \|f\|^2$$

if and only if f *lies in the closed linear span of* (ϕ_n).

This equality is often called the *Parseval relation*; note that it implies $c_n \to 0$ as $n \to \infty$.

The following theorem is a sort of converse to the Bessel inequality and is sometimes called the Riesz–Fischer theorem. A stronger form is given in the next section.

THEOREM   *Let (ϕ_n) be an ON sequence in H and (a_n) a sequence of scalars such that $\Sigma |a_n|^2 < \infty$. Then $\Sigma a_n \phi_n$ converges to some f in H.*
   Proof   Put

$$S_N = \sum_{n=1}^{N} a_n \phi_n.$$

Then

$$\|S_N - S_M\|^2 = \left\| \sum_{n=N+1}^{M} a_n \phi_n \right\|^2 = \sum_{n=N+1}^{M} |a_n|^2$$

upon expanding the norm as an inner product and using orthonormality. But since $\Sigma |a_n|^2$ is convergent, its partial sums form a Cauchy sequence of reals, hence (S_N) is a Cauchy sequence and the result follows by completeness of H.

1.3.2   Complete and total sets   From the results of the previous section an important aspect of ON sets is beginning to emerge: it is the possibility of expanding a member of a space into a Fourier series with convergence in the strong sense. In this context it will be an important property of a sequence, ON or not, if we can assert that its closed linear span is the whole space. This property is called 'totality' (see definition below). A closely related property (in Hilbert space entirely equivalent) is that of 'completeness'. Before proceeding to the definitions and the proof of equivalence we must point out that the word 'completeness' has already been assigned a technical meaning; the present usage is fairly standard, however, and we shall not depart from it. Nevertheless, to add to the confusion some authors use 'completeness' for the property we shall call 'totality'; others have used 'closure', now mercifully outdated. Furthermore, the words 'maximal', 'minimal', 'fundamental' and 'basic' are also to be found in the literature associated with these or closely related properties. Forewarned is forearmed!

DEFINITION   A sequence $(\phi_n)_{n \in J}$ in V (we require this definition only for V as described in §1.1.2) is called *total* in V if the class of all finite linear combinations $\Sigma a_n \phi_n$ is dense in V. This is the same as saying that $[\phi_n] = V$.

DEFINITION A sequence $(\phi_n)_{n \in J}$ in a Hilbert space H is called *complete* if the only element of H which is orthogonal to every ϕ_n is the null element. That is,

$$(f, \phi_n) = 0 \quad (n \in J) \quad \Rightarrow f = \theta.$$

Note that these definitions apply to any sequences in H, orthogonal or not. The completeness property can also be formulated for L^p spaces (see problem 1.8). We also note that *if (ϕ_n) is an ON sequence, then it is total in H if and only if Parseval's relation is satisfied for every f in H.*

THEOREM *If (ϕ_n) is any sequence in H, orthogonal or not, then it is complete in H if and only if it is total in H.*

We give the proof of the case where (ϕ_n) is an ON sequence; the reader will have no difficulty in extending the proof to the case of non-orthogonal sequences by appeal to the Gram–Schmidt process (Appendix 1,7).

Proof (i) Completeness implies totality. Let $f \in H$. Now $c_n = (f, \phi_n)$ are, by Bessel's inequality, such that $\Sigma |c_n|^2 < \infty$. Hence by the last theorem of the previous section there exists g in H such that

$$\left\| g - \sum_{n=1}^{N} c_n \phi_n \right\| \to 0 \quad (N \to \infty).$$

Since g determines its coefficients uniquely as Fourier coefficients, we find that the Fourier coefficients of f are equal to those of g and hence those of $f - g$ are all zero. Then by completeness, $f = g$. Hence every f of H has a Fourier expansion in the set $\{\phi_n\}$ which shows that (ϕ_n) is total in H.

As we have already remarked, the fact that every f of H has a Fourier expansion in the set $\{\phi_n\}$ is the key to the importance of complete orthogonal sequences in Hilbert space. It generalises the idea of representing a vector of a finite dimensional vector space in terms of its components with respect to a basis of unit

vectors. If the sequence (ϕ_n) is not orthogonal this Fourier
expansion property may fail to hold, even if (ϕ_n) is complete,
and it is this consideration which is behind the definition of a
basis for Hilbert space (see definition, § 1.4, and problem 1.14).

Proof (ii)   Totality implies completeness. Let $c_n = (f, \phi_n) = 0$
for every $n \in J$ and (ϕ_n) be total. Then f has a Fourier expansion
in the set $\{\phi_n\}$ with zero coefficients, and Parseval's theorem
yields $f = \theta$. This completes the proof.

Henceforth we use the equivalence asserted by this theorem
without further mention. We use the letters CON to denote a
complete/total ON sequence in H. We are now in a position to
strengthen the last theorem of § 1.3.1. See also problem 1.9.

THEOREM (F. Riesz and E. Fischer)   *Let (ϕ_n) be an ON
sequence in a Hilbert space H and (a_n) a sequence of scalars. Then*

(i) *In order that there exist $f \in H$ such that $\Sigma a_n \phi_n$ converges to f
it is necessary and sufficient that $\Sigma |a_n|^2 < \infty$. Under either
condition, $a_n = (f, \phi_n)$.*

(ii) *If (ϕ_n) is complete then there cannot exist $g \in H$, distinct
from f for which $a_n = (g, \phi_n)$ for every n.*

Proof.   Part (i) summarises results already to hand from the
previous section. For part (ii) we merely note that if $a_n = (g, \phi_n)$
for every n, then $0 = (g, \phi_n) - (f, \phi_n) = (g - f, \phi_n)$ and by completeness of (ϕ_n), $f - g = \theta$, that is, $f = g$ and the proof is complete.

We have established the equivalence of completeness and
totality in Hilbert space. For ON sequences there are five
equivalent properties and a list of these follows; we have
established all except the last whose proof is left as an exercise
for the reader.

Let (ϕ_n) be an ON sequence in H. Then the following statements
are equivalent.

(1) (ϕ_n) is complete.

(2) (ϕ_n) is total.

(3) The Parseval relation is satisfied for every $f \in H$.

(4) For every $f \in H, f = \Sigma c_n \phi_n$ with $c_n = (f, \phi_n)$.

(5) A more general Parseval relation is satisfied: for every f and g in H,
$$(f, g) = \Sigma c_n \bar{d}_n,$$

where c_n and d_n are the Fourier coefficients of f and g respectively (see problem 1.9).

In view of the importance of complete orthonormal sets, the following theorem is one of the fundamental structure theorems for Hilbert space.

THEOREM *Every Hilbert space contains a complete orthonormal sequence.*

Proof Let O denote the collection of all ON sequences in H. O is not empty; a single normalised element is an example of an ON sequence, and if H contains linearly independent elements these can be orthogonalised by the Gram–Schmidt process. Then inclusion, in the set theoretic sense, is a partial order (see Appendix 1,6(*a*)) on O. Let L be a linearly ordered subset of O and M the union of all members of L. Then $M \in O$ and clearly M is an upper bound for L. Zorn's lemma (Appendix 1,6(*b*)) applies and we find that O has a maximal element; this is a maximal ON sequence $(\phi_n)_{n \in J}$ in the sense that it is not a proper subsequence of any other ON sequence. As such, it must be complete, for if f is orthogonal to every ϕ_n and f is not the null element then the maximality of (ϕ_n) is contradicted; this establishes the result.

Problems

1.8 Let (a, b) be a finite interval and (ϕ_n) a sequence in $L^q(a, b)$, then (ϕ_n) is said to be complete on $L^p(a, b)$, $p + q = pq$, if, for $f \in L^p(a, b)$
$$\int_a^b f\phi_n = 0 \text{ for every } n$$
implies f null. Show that

(i) If (ϕ_n) is total in $L^p(a, b)$, $1 \leqslant p \leqslant \infty$, then it is complete on $L^q(a,b)$. (The converse is also true, see Kaczmarz and Steinhaus (1935) p. 200.)

(ii) Let $1 \leqslant p < p' \leqslant \infty$. Show that if (ϕ_n) is total in $L^{p'}(a,b)$ then it is total in $L^p(a,b)$, whilst if (ϕ_n) is complete on $L^p(a,b)$ then it is complete on $L^{p'}(a,b)$. Show that completeness on $L(a,b)$ implies completeness on, and totality in $L^p(a,b)$, $p \geqslant 1$.

1.9 Let (ϕ_n) be orthogonal but not complete in a Hilbert space H. Show that there exist f, $g \in H$ with $(f,\phi_n) = (g,\phi_n)$ for every n, $f \neq g$ (cf. the Riesz–Fischer theorem).

1.10 Show that item 5 does belong in the list of equivalent properties of ON sequences on p. 17. Show that the convergence in this relation is absolute.

1.11 Show that any two members of a CON sequence in H are distant $\sqrt{2}$ apart. Hence if H is separable, every CON sequence in H is countable.

1.12 Show that any two CON sequences in separable Hilbert space have the same cardinality.

1.13 Show that two Hilbert spaces are isomorphic if and only if they have the property that the cardinality common to all the CON sequences in one of them is equal to that of the other.

DEFINITION The cardinality common to all CON sequences in separable Hilbert space (see problem 1.12) is called the *dimension* of the space (sometimes the *Hilbert dimension*).

There are examples of Hilbert spaces whose dimension is uncountable, the non-separable Hilbert spaces; they occur, for example, in the theory of almost periodic functions. However we shall assume throughout that all our Hilbert spaces are separable, of countably infinite dimension and that their field of scalars is the complex number field.

1.4 Biorthogonal systems in Hilbert space

In this section we introduce a class of sequences in Hilbert space which are not orthogonal but which have an associated 'biorthogonal' sequence according to the following definition.

DEFINITION Let (ϕ_n) and (ϕ_n^*) be two sequences in a Hilbert space H. If $(\phi_n, \phi_m^*) = 0$, $n \neq m$, then each sequence is said to have the other as a *biorthogonal sequence*, and if $(\phi_n, \phi_n^*) = 1$, the collection $\{\phi_n, \phi_n^*\}$ is said to be *normal*. More compactly, if $(\phi_n, \phi_m^*) = \delta_{nm}$, then $\{\phi_n, \phi_n^*\}$ is called a *biorthonormal system* (BON system for short).

An objection might be raised here. We have the Gram–Schmidt orthogonalisation process, so why do we not just orthogonalise (ϕ_n) and obviate the necessity of dealing with non-orthogonal sequences altogether? Actually there are various reasons for not orthogonalising. One of these is the intrinsically complicated nature of the Gram–Schmidt process; the orthogonalised sequence may become quite unmanageable. (This is by no means always the case however; see §2.1.) A more important consideration is that a non-orthogonal sequence may arise in some context where it has a particular property which would be destroyed by orthogonalisation; the upshot is that we must be able to deal directly with non-orthogonal sequences.

Two points about the definition of a BON system must be carefully noted. One is that it does not follow from the definition that either (ϕ_n) or (ϕ_n^*) is normalised in the sense that each member has unit norm; indeed, if they both are, then a BON system forming a basis (see below) degenerates to a CON sequence (see Marti (1969) p. 81). The second is that if (ϕ_n) is complete then (ϕ_n^*) is not necessarily complete; the following example illustrates the point.

EXAMPLE (Kaczmarz and Steinhaus (1935) p. 262) Let $(\psi_n)_{n=1}^{\infty}$ be a CON sequence in H. Put

$$\phi_n = \psi_1 + \psi_{n+1} \quad (n = 1, 2, \ldots)$$

and

$$\phi_n^* = \psi_{n+1} \quad (n = 1, 2, \ldots).$$

Then $\{\phi_n, \phi_n^*\}$ is obviously a BON system, but as we shall see (ϕ_n) is complete but (ϕ_n^*) is not. Now (ϕ_n) is a complete sequence, for, if $(f, \phi_n) = 0$ for every n and some $f \in H$, then

$$(f, \psi_1) = -(f, \psi_{n+1})$$

for every n. But $\lim_{n \to \infty} (f, \psi_{n+1}) = 0$ (see the remark following Parseval's theorem), hence $(f, \psi_1) = 0$; it follows that $(f, \psi_n) = 0$ for every n. But (ψ_n) is complete, therefore $f = \theta$ and hence (ϕ_n) is complete. On the other hand, (ϕ_n^*) is not complete for ψ_1 is orthogonal to every ϕ_n^*, and this completes the demonstration.

We have seen that the expansion theory for CON sequences in Hilbert space is very satisfactory, whereas if a sequence is complete but not orthogonal the situation is far less satisfactory. Indeed, if (ϕ_n) is total but not orthogonal the most we can say about an f in H is that, given $\epsilon > 0$, there is some finite sequence of coefficients (a_n), such that $\|f - \Sigma a_n \phi_n\| < \epsilon$; if a new ϵ is given, the whole set of coefficients may have to be changed. However, there are non-orthogonal sequences with a fruitful expansion theory and this leads us to the idea of a 'basis'. The two theorems which follow the definition are fundamental structure theorems for Hilbert space.

DEFINITION   The sequence (ϕ_n) in H is called a *basis* for H if for every $f \in H$ there exists a unique sequence of scalars (a_n) such that $\qquad f = \Sigma a_n \phi_n$, in the strong sense.

In particular, a CON sequence is a basis for Hilbert space.

Again it will be convenient to be rather free with the terminology and speak for example of a *'set* of functions forming a basis...', it being understood that the set constitutes the range of the sequence under consideration.

THEOREM   *The sequence (ϕ_n) is a basis for H if and only if it has a unique biorthogonal sequence (ϕ_n^*) and $\Sigma(f, \phi_n^*) \phi_n$ converges to f for every f in H.*

Proof (i) 'If' We show first that (ϕ_n) is a basis if it has a biorthogonal set for which $\Sigma(f, \phi_n^*) \phi_n$ converges to f for every $f \in H$. We may assume that $\{\phi_n, \phi_n^*\}$ is normal as well as biorthogonal. We need only prove that the coefficients $c_n = (f, \phi_n^*)$ are unique. Indeed, if (b_n) is any sequence of coefficients for which $\Sigma b_n \phi_n$ converges to f, then

$$|(f, \phi_n^*) - b_n| = \left| \left(f - \sum_{m=1}^{N} b_m \phi_m, \phi_n^* \right) \right|, \quad N \geqslant n$$

$$\leqslant \left\| f - \sum_{m=1}^{N} b_m \phi_m \right\| \|\phi_n^*\|$$

$$\to 0 \text{ as } N \to \infty, \text{ for every } n,$$

and this completes the first part.

The converse is the deeper part of the theorem. For this we shall need the idea of a 'coefficient functional'.

DEFINITION   Let (ϕ_n) be a basis for H, and (a_n) the sequence of coefficients uniquely determined by $f \in H$. Then the nth *coefficient functional* with respect to (ϕ_n) is defined by

$$\alpha_n(f) = a_n.$$

LEMMA   *For every n, α_n is a bounded linear functional.*

Proof The reader will easily verify the linearity. We also leave it to the reader to verify that the collection of all sequences of scalars $(a_n) = A$ for which $\Sigma a_n \phi_n$ converges is a vector space. We denote this space by S; then we have

$$\sup_N \left\| \sum_{n=1}^{N} a_n \phi_n \right\| < \infty \quad (A \in S).$$

(i) The first part of the proof consists of showing that S is complete, and hence that it is a Banach space, in the norm defined by

$$\|A\|_S = \sup_N \left\| \sum_{n=1}^{N} a_n \phi_n \right\|.$$

2425

To show this let $(A^{(p)})$ be a Cauchy sequence in S, i.e. given $\epsilon > 0$ there exists M such that

$$\|A^{(p)} - A^{(q)}\|_S = \sup_N \left\| \sum_{n=1}^N (a_n^{(p)} - a_n^{(q)}) \phi_n \right\|$$

$$\leqslant \epsilon \quad (p, q > M).$$

We are going to show, as an intermediate step, that for each integer i, $(a_i^{(p)})$ is a Cauchy sequence of scalars; as such it will be convergent, to a limit to be denoted by a_i. Now

$$\|(a_i^{(p)} - a_i^{(q)}) \phi_i\| = \left\| \left(\sum_{n=1}^i - \sum_{n=1}^{i-1} \right) (a_n^{(p)} - a_n^{(q)}) \phi_n \right\|$$

$$\leqslant 2\epsilon \text{ for every } i \quad (p, q > M).$$

Therefore $|a_i^{(p)} - a_i^{(q)}| < \epsilon / \|\phi_i\|$, which gives the required Cauchy property. We have

$$\left\| \sum_{n=1}^i (a_n^{(p)} - a_n^{(q)}) \phi_n \right\| < \epsilon \text{ for every } i \quad (p, q > M).$$

Therefore letting $q \to \infty$ we have

$$\left\| \sum_{n=1}^i (a_n^{(p)} - a_n) \phi_n \right\| \leqslant \epsilon \text{ for every } i \quad (p > M).$$

We want to show that the sequence $(a_n) = A$ that we have constructed lies in S. To do this we must show that the sequence (Σ_N) of partial sums $\left(\sum_{n=1}^N a_n \phi_n \right)$ converges in H; by the completeness of H it will be sufficient to show that (Σ_N) is Cauchy. Consequently

$$\left\| \sum_{n=i+1}^{i+k} a_n \phi_n \right\|$$

$$= \left\| \sum_{n=1}^{i+k} (a_n \phi_n - a_n^{(p)} \phi_n) - \sum_{n=1}^i (a_n - a_n^{(p)}) \phi_n + \sum_{n=i+1}^{i+k} a_n^{(p)} \phi_n \right\|$$

$$\leqslant 2\epsilon + \left\| \sum_{n=i+1}^{i+k} a_n^{(p)} \phi_n \right\|.$$

But $\Sigma a_n^{(p)} \phi_n$ is convergent, and so its partial sums form a

Cauchy sequence. This shows that (Σ_N) is Cauchy. S is indeed a Banach space, for $(A^{(p)})$ converges because

$$\|A^{(p)} - A\|_S = \sup_N \left\| \sum_{n=1}^{N} (a_n^{(p)} - a_n)\, \phi_n \right\|$$

$$< \epsilon.$$

In the second part of the proof the required boundedness of α_n will follow by applying the 'bounded inverse' theorem (Appendix 1,5) to the transformation $T: S \to H$ under which $A \to f = \Sigma\, a_n\, \phi_n$.

We note first that, since each f has an expansion in the set $\{\phi_n\}$ with unique coefficients, T is one-to-one and onto; thus T^{-1} exists. Furthermore, T is bounded, for

$$\|T(A)\| = \|\Sigma\, a_n\, \phi_n\|$$

$$\leqslant \sup_N \left\| \sum_{n=1}^{N} a_n\, \phi_n \right\|$$

$$= \|A\|_S.$$

Thus T^{-1} is bounded, by the bounded inverse theorem. Hence

$$|a_i|\, \|\phi_i\| = \left\| \left(\sum_{n=1}^{i} - \sum_{n=1}^{i-1} \right) (a_n\, \phi_n) \right\|$$

$$\leqslant 2\|A\|_S$$

$$= 2\|T^{-1}f\|_S$$

$$\leqslant 2\|T^{-1}\|\, \|f\|.$$

Therefore $|\alpha_i(f)| = |a_i|$

$$\leqslant K_i\|f\| \text{ for every } i$$

by uniqueness of the a_is. This completes the proof.

We can now complete the proof of the theorem.

Proof (ii) 'Only if' Since the coefficient functionals are bounded, the Riesz–Fréchet representation theorem (Appendix 1,4) shows that for each n there exists a uniquely defined $\phi_n^* \in H$

such that $\alpha_n(f) = (f, \phi_n^*)$. Hence $\Sigma(f, \phi_n^*)\,\phi_n$ converges to f for every f in H. In particular,

$$\phi_k = \Sigma(\phi_k, \phi_n^*)\,\phi_n$$

and by uniqueness of the coefficients we must have

$$(\phi_k, \phi_n^*) = \delta_{kn}.$$

This completes the proof.

The final theorem of this chapter underlines how strong the property of being a basis really is, for as we have seen in the example of this section the corresponding result for sequences which are merely complete may fail to hold.

THEOREM   *Let $\{\phi_n, \phi_n^*\}$ be a BON system in H. Then if (ϕ_n) is a basis, so is (ϕ_n^*).*

Proof   The proof is in three parts.
(i) Let f and g lie in H, put $b_n = (g, \phi_n)$ and let (a_n) be a sequence of scalars such that $\Sigma a_n \phi_n$ converges weakly to f. Then $\Sigma \bar{a}_n b_n = (g,f)$, for

$$\sum_{n=1}^{N} \bar{a}_n b_n = \sum_{n=1}^{N} \bar{a}_n(g, \phi_n) = \left(g, \sum_{n=1}^{N} a_n \phi_n\right).$$

If we now let $N \to \infty$ the required result follows.

(ii) Put $a_n = (f, \phi_n^*)$. If $\Sigma a_n \phi_n$ converges weakly to f for every f in H, then $\Sigma b_n \phi_n^*$ converges weakly to g for every g in H, since for every f in H we have

$$\lim_{N \to \infty} \left(f, \sum_{n=1}^{N} b_n \phi_n^*\right) = \lim_{N \to \infty} \sum_{n=1}^{N} \bar{b}_n a_n = (f, g),$$

by part (i).

(iii) With the notations of the previous part, we now show that, if $\Sigma a_n \phi_n$ converges weakly to f for every f in H, then $\Sigma b_n \phi_n^*$ converges strongly to g for every g in H.

We observe that $S_m(g) = \sum_{n=1}^{m} b_n \phi_n^*$ is, for every m, a linear operator on H; we shall apply the uniform boundedness principle

(Appendix 1,1) to the sequence $(S_m(g))$. Now, each $S_m(g)$ is bounded, for

$$\|S_m(g)\| = \left\| \sum_{n=1}^{m} b_n \phi_n^* \right\|$$

$$\leqslant \sum_{n=1}^{m} |b_n|\, \|\phi_n^*\|$$

$$\leqslant \sum_{n=1}^{m} \|g\|\, \|\phi_n\|\, \|\phi_n^*\|$$

$$= \|g\| C_m.$$

Further, for every g there exists a constant B_g such that $\|S_m(g)\| \leqslant B_g$, since g is the weak limit of $S_m(g)$ (Appendix 1,2). Hence the uniform boundedness principle applies, so that there exists a constant A such that

$$\|S_m(g)\| \leqslant A\|g\| \quad (g \in H).$$

Also g lies in the closed linear span of $\{S_m(g)\}$; this too follows from the fact that g is the weak limit of $S_m(g)$ (Appendix 1, 2) Therefore, given $\epsilon > 0$, there exist coefficients a_{mn} and an integer M_g such that

$$\left\| g - \sum_{n=1}^{m} a_{mn} S_n(g) \right\| < \epsilon \quad (n > M_g).$$

Now let σ_m denote $\sum_{n=1}^{m} a_{mn} S_n(g)$; we leave it to the reader to verify that $S_m(\sigma_m) = \sigma_m$. We now use the various results of this third part of the proof to obtain

$$\|S_m(g) - \sigma_m\| = \|S_m(g - \sigma_m)\|$$

$$\leqslant A\|g - \sigma_m\|$$

$$< A\epsilon \quad (m > M_g).$$

Finally, $\|g - S_m(g)\| = \|g - \sigma_m - S_m(g) + \sigma_m\|$

$$\leqslant \|g - \sigma_m\| + \|S_m(g) - \sigma_m\|$$

$$< \epsilon + A\epsilon \quad (m > M_g).$$

This completes the third part, and the result of the theorem now follows, for, if (ϕ_n) is a basis, then by the previous theorem $\Sigma a_n \phi_n$ converges strongly (hence weakly) to f for every f in H, so that $\Sigma b_n \phi_n^*$ converges strongly to g for every g in H, and again by the previous theorem (ϕ_n^*) is a basis.

Some further remarks can be made following this theorem and its proof:

(1) Let (ϕ_n) be a basis for H and for f and g in H put $a_n = (f, \phi_n^*)$ and $b_n = (g, \phi_n)$. Since strong convergence implies weak convergence, part (i) of the above proof yields the more *general Parseval relation*

$$\Sigma a_n \bar{b}_n = (f, g).$$

(2) With the notations as above, a sequence (ϕ_n) is called a *weak basis* for H if $\Sigma a_n \phi_n$ converges weakly to f for every f in H. Then (ϕ_n) *is a basis if and only if it is a weak basis*. It is only necessary to show that a weak basis is a strong basis; but this follows immediately from part (iii) of the previous proof.

(3) There is an obvious duality, as far as basis properties are concerned, between (ϕ_n) and (ϕ_n^*); thus if either is a basis there are two series expansions $\Sigma a_n \phi_n$ and $\Sigma a_n^* \phi_n^*$ for every f in H, where $a_n^* = (f, \phi_n)$.

(4) The concept of basis can be extended to Banach space; here the coefficients in the series are linear functionals in the dual space B^*. The reader will find it instructive to formulate the appropriate definition, and to extend the two previous theorems and their proofs to the case of Banach space (see Marti (1969) pp. 31–3).

Problem

1.14 Show that every Hilbert space contains a complete sequence which is not a basis.

1.5   Postscript to chapter 1

The reader may wonder if the series we have been discussing are independent of the order in which they are written. The following definitions and remarks will help to clarify the situation, but we must refer the reader elsewhere, e.g. Marti (1969), for further details.

DEFINITIONS   An *unconditional basis* (ϕ_n) for a Hilbert or, indeed, a Banach space is a basis such that, for every f in the space, $\Sigma\, \alpha_n(f)\, \phi_n$ converges, regardless of the order in which the terms are written; here, α_n are the coefficient functionals associated with (ϕ_n). A *conditional basis* is one which is not unconditional. A basis (ϕ_n) is called an *absolute basis* if, for every f in the space, $\Sigma\, \|\alpha_n(f)\, \phi_n\|$ converges.

An absolute basis is unconditional. Not all bases for a Hilbert space H are unconditional and hence not all are absolute; however, a normal basis for H is unconditional if and only if (with the notations of the previous theorem) both $\Sigma |a_n|^2$ and $\Sigma |a_n^*|^2$ converge for every f in H; in particular every CON sequence in H is an unconditional basis.

As we have seen, every Hilbert space possesses a CON sequence, that is, every Hilbert space has a basis; Banach (1932, p. 111) observed '*on ne sait pas si tout espace du type (B) séparable admet une base*'. This 'basis problem', as it became known, remained one of the famous unsolved problems of mathematics until, four decades later, it was shown by Per Enflo (1973) that there exists a reflexive separable Banach space that fails to have the approximation property; this implies (see e.g. Singer (1970) p. 170) that it has no basis.

Nevertheless, many well-known Banach spaces have long been known to possess a basis; for example, the Haar system (see § 2.5) is a basis for $L^p(0, 1)$, $1 \leqslant p < \infty$ (unconditional if $p > 1$, conditional if $p = 1$), and the Schauder system (Marti (1969) p. 49) is a conditional basis for $C(0, 1)$. The trigonometrical system (see § 2.1) is a basis for $L^p(-\pi, \pi)$, $1 < p < \infty$, which is conditional for $p \neq 2$. The trigonometrical system is *not* a basis for $L(-\pi, \pi)$, indeed it is known that there are functions whose Fourier series diverge in the L norm.

2. Orthogonal Sequences

We dealt with the theoretical aspects of bases in the first chapter, and now the reader will be anxious to see some concrete examples. Consequently this chapter and the next will be devoted to developing various methods for demonstrating completeness and basis properties, and to the application of these methods to particular examples. In the present chapter the methods are mostly designed to treat orthogonal sequences, and we leave a more detailed account of non-orthogonal sequences to Chapter 3.

2.1 Complete sequences of polynomials

Many of the standard L^2 spaces, for example those taken over finite or semi-infinite intervals of $\mathbb{R}$, or over $\mathbb{R}$ itself, have bases consisting of sequences of polynomials (p_n) orthogonal with respect to a weight function w. This means that we have, for some measurable subset E of $\mathbb{R}$,

$$\int_E p_n p_m w = \delta_{nm},$$

and that we can consider either (p_n) to be a CON sequence in $L^2(E, w)$, or $(\sqrt{(w)}\,p_n)$ to be a CON sequence in $L^2(E)$. Such sequences arise by orthogonalising the set of powers

$$\{x^n : n = 0, 1, \ldots\}$$

with respect to w by the Gram–Schmidt orthogonalisation process. The completeness of the resulting set is not at once evident; consequently, we shall give a theorem in this section which guarantees completeness for suitably chosen weight functions. The uniqueness of the resulting sequence is guaranteed by the Gram–Schmidt process (see p. 117). The foregoing facts serve to characterise certain sets of polynomials, in that for certain L^2 spaces they constitute the only possible orthogonal polynomial basis containing a polynomial of each degree. Some examples are: the orthogonal polynomials of Jacobi in $L^2((-1,1)$, $(1-x)^\alpha (1+x)^\beta)$, α, $\beta > -1$, which include several important special cases (see Appendix 2,5), those of Laguerre in $L^2(\mathbb{R}^+, e^{-x})$, and those of Hermite in $L^2(\mathbb{R}, e^{-x^2})$.

In order to approach the completeness problem, let us restate the result derived at the end of § 1.2 in terms of completeness: *the set of powers* $\{x^n : n = 0, 1, \ldots\}$ *forms a complete sequence in* $L^2(a,b)$ *for any finite interval* (a,b); *indeed it is complete on* $L^p(a,b)$, $1 \leqslant p < \infty$.

The completeness theorem for polynomials is a consequence of the following.

LEMMA *Let* w *be integrable over* $\mathbb{R}$ *and such that* $w(x) > 0$ *a.e. let* f *lie in* $L(\mathbb{R}, w)$ *and suppose that*

$$\int_\mathbb{R} f(x)\, e^{itx} w(x)\, dx = 0 \quad (t \in \mathbb{R}).$$

Then f *is null.*

Proof For every trigonometrical polynomial

$$t_n(x) = \sum_{|k| \leqslant n} c_k e^{2\pi i k x/\omega}$$

of period ω we have, by hypothesis,

$$\int_\mathbb{R} t_n(x) f(x)\, w(x)\, dx = 0.$$

The proof consists of extending this formula, in four stages, from trigonometrical polynomials to bounded measurable functions on $\mathbb{R}$. We can then apply it to the function sgn f and obtain

$$\int_\mathbb{R} \operatorname{sgn} f(x) f(x)\, w(x)\, dx = \int_\mathbb{R} |f(x)|\, w(x)\, dx = 0,$$

from which we conclude that f is null.

In the first stage we extend the formula to continuous,

ω-periodic functions using the second form of the Weierstrass theorem (Appendix 1,3(b)). For, given such a function c, and an $\epsilon > 0$, we can find a trigonometrical polynomial t_n of period ω such that $|c - t_n| < \epsilon$ for each x in the period interval and thus each x in $\mathbb{R}$. Then a simple calculation yields

$$\int_{\mathbb{R}} f(x)\, c(x)\, w(x)\, dx = 0$$

for every continuous ω-periodic c on $\mathbb{R}$.

The second stage is to extend this formula to functions which are continuous on $\mathbb{R}$ and which vanish outside a finite interval (a, b), or, as we shall say, *continuous functions with support on* (a, b). Given such a function h, pick ω so that $(a, b) \subset (-\tfrac{1}{2}\omega, \tfrac{1}{2}\omega)$ and put $h_\omega = h$ on $(-\tfrac{1}{2}\omega, \tfrac{1}{2}\omega)$ and extend h_ω to $\mathbb{R}$ by periodicity. Then $\lim_{\omega \to \infty} h_\omega = h$. Further, $|h_\omega| \leqslant \max |h|$, so that

$$\int_{\mathbb{R}} h(x) f(x)\, w(x)\, dx = \lim_{\omega \to \infty} \int_{\mathbb{R}} h_\omega(x) f(x)\, w(x)\, dx$$

by the Lebesgue dominated convergence theorem. We now have the extension

$$\int_{\mathbb{R}} h(x) f(x)\, w(x)\, dx = 0$$

for every continuous h with support on a finite interval.

The third extension is to step functions s with support on a finite interval I. By an obvious construction there exists a sequence (h_n) of continuous functions with support on I such that $s = \lim h_n$ everywhere on I except at the discontinuity points of s, that is, a.e. on I. Further, $|h_n| \leqslant |s|$ for every n a.e. on I. Then by dominated convergence

$$\int_I sfw = \lim \int_I h_n fw = 0.$$

The final extension is to bounded measurable functions g on $\mathbb{R}$. Define g_N to be equal to g on $(-N, N)$ and to vanish outside $(-N, N)$.

It is known that one can find a sequence (s_{Nn}) of step functions with support on $(-N, N)$ converging a.e. to g_N and such that the bounds of each s_{Nn} are equal to those of g_N. Then (s_{NN}) is a sequence of step functions with $g = \lim\limits_{N \to \infty} s_{NN}$ a.e. and, by dominated convergence,

$$\int_{\mathbb{R}} gfw = \lim_{N \to \infty} \int_{\mathbb{R}} s_{NN} fw = 0.$$

This completes the proof of the lemma.

DEFINITION A set of polynomials $\{p_n : n = 0, 1, 2, ...\}$ is called *simple* if, for every n, p_n is of degree n.

THEOREM (Completeness theorem for polynomials) *Let (a, b) be a finite or infinite interval of $\mathbb{R}$ and w a non-negative measurable weight function on (a, b) such that there exists $r > 0$ for which*

$$\int_a^b e^{r|x|} w(x)\, dx < \infty.$$

Then any simple set of polynomials $\{p_n : n = 0, 1, ...\}$ is complete in $L^2((a, b), w)$.

Proof It is necessary to check that under the hypotheses each x^n is indeed a member of $L^2((a, b), w)$: this follows at once, since for every n and for every $r > 0$ there exists a constant A for which $|x|^{2n} < Ae^{r|x|}$ for all sufficiently large x. Obviously w is integrable over (a, b).

Assume that there is an $f \in L^2((a, b), w)$ which is not null and which is orthogonal to every p_n, or, equivalently, for which

$$\int_a^b x^n f(x)\, w(x)\, dx = 0, \quad n = 0, 1, ...$$

Put

$$F(z) = \int_a^b f(x)\, e^{izx}\, w(x)\, dx, \quad z = x + iy,$$

then Schwarz's inequality shows that

$$|F(z)|^2 \leqslant \int_a^b |f(x)|^2 w(x)\, dx \int_a^b e^{-2yx}\, w(x)\, dx < \infty, |y| < r/2.$$

Similarly,

$$|F'(z)|^2 \leqslant \int_a^b |f(x)|^2 w(x)\, dx \int_a^b |x|^2\, e^{-2yx}\, w(x)\, dx < \infty$$

for sufficiently small $|y|$, certainly for $|y| < r/4$. Thus F is a regular function for $|y| < r/4$. On the other hand we can expand F as a power series convergent for sufficiently small $|z|$ by writing

$$F(z) = \int_a^b \sum_{k=0}^\infty \frac{(izx)^k}{k!} f(x)\, w(x)\, dx$$

and integrating term by term. This is permissible by Beppo Levi's theorem, since

$$\int_a^b \sum_{k=0}^\infty \frac{|zx|^k}{k!} f(x)\, w(x)\, dx$$

$$= \int_a^b e^{|zx|} f(x)\, w(x)\, dx$$

$$\leqslant \left\{ \int_a^b f^2(x)\, w(x)\, dx \int_a^b e^{2|zx|} w(x)\, dx \right\}^{\frac{1}{2}}$$

$$< \infty \text{ provided } 2|z| < r.$$

That is, we have

$$F(z) = \sum_{k=0}^\infty \frac{z^k}{k!} \left\{ \int_a^b x^k f(x)\, w(x)\, dx \right\}, \quad |z| < r/2.$$

Hence $F(z) = 0$ for $|z| < r/2$, therefore $F(z) \equiv 0$ for $|y| < r/4$ by the identity theorem for analytic functions. In particular

$$F(t) = \int_a^b e^{itx} f(x)\, w(x)\, dx = 0 \text{ for every } t \in \mathbb{R}.$$

If (a, b) is not all of $\mathbb{R}$, w and f can be extended to $\mathbb{R}$ as functions with support on (a, b). As such, $w \in L(\mathbb{R})$ and by Schwarz's inequality $f \in L(\mathbb{R}, w)$. Then the previous Lemma applies and f is null; this contradiction completes the proof.

Note that the simple property of the set $\{p_n\}$ was used at the beginning of the proof to assert that if f is orthogonal to every p_n then it is orthogonal to every x^n. This may not be the case if $\{p_n\}$ is not simple, see for example problem 2.3 Many of the sets of polynomials arising in the special function theory are simple sets, and the completeness theorem applies to them whether or not they are orthogonal. Some examples are the polynomials of Bernoulli and of Euler, and many sets of hypergeometric type including the very general polynomials of Sister Celine (see Rainville (1963) for the definitions and many more examples).

Problems

2.1 Show that the completeness theorem for polynomials applies to the polynomials of Jacobi, Laguerre and Hermite in their respective L^2 spaces (see the first paragraph of §2.1). If α and β are only required to be positive, prove the completeness of the Jacobi polynomials by another method. Show also that it applies to the sequences obtained by orthonormalising the powers (x^n) over $\mathbb{R}$ with respect to the weights $(1+x^{2k})^\alpha e^{-x^{2k}}$, k a natural number, $\alpha \geqslant 0$ (such sequences arise in approximation theory (Nevai (1973), and evidently generalise the Hermite polynomials).

2.2 Show that the completeness theorem for polynomials holds, without further assumptions, for $L^p((a,b),w)$, $1 < p < \infty$.

2.3 Find an example of a set of polynomials $\{p_n : n = 0, 1, ...\}$ on (a,b) with the property that every power x^k, $k = 0, 1, ...$, occurs as a term in at least one of the p_n, but that (p_n) fails to be complete in $L^2(a,b)$.

2.4 Find an orthogonal polynomial basis for $L^2(-1,1)$ other than the Legendre polynomial basis. Hint: refer to the theorem of Müntz, p. 95.

2.2 The Vitali completeness criterion

Let us define a sequence of step functions $(\chi_r(x))$ on the finite interval (a,b) by the formula

$$\chi_r(x) = \begin{cases} 1, & a \leqslant x \leqslant r \\ 0, & r < x \leqslant b, \end{cases}$$

where r may take any value between a and b. We show that (χ_r) is complete in $L^2(a,b)$. To see this let f be any member of $L^2(a,b)$ and suppose that

$$\int_a^b \chi_r(x)f(x)\,dx = \int_a^r f(x)\,dx = 0$$

for every $r \in (a, b)$. That is,

$$F(r) = \int_a^r f(x)\,dx \equiv 0.$$

Now f is integrable on (a, b) (since $L^2(a, b) \subset L^1(a, b)$), so that $f = F'$ a.e. according to the well-known Lebesgue theory. But F' is null, thus the definition of completeness is satisfied.

The following modifications to the sequence (χ_r) are useful; the proofs of 2 and 3 are left as exercises for the reader.

(1) The set of functions defined by

$$\xi_r(x) = \begin{cases} 1, & a \leqslant x \leqslant r \\ -1, & r < x \leqslant b, \end{cases}$$

where r may take any value between a and b forms a complete sequence in $L^2(a, b)$. For

$$\int_a^b \xi_r f = 2\int_a^r f - \int_a^b f,$$

hence, if

$$F(r) = \int_a^r f - \tfrac{1}{2}\int_a^b f,$$

then $F'(r) = f$ a.e. and the proof is completed as before.

(2) Let (a, b) be a finite or infinite interval of $\mathbb{R}$, and

$$g \in L^2((a, b), w),$$

where w is a positive continuous weight function. Then both $(g(x)\,\chi_r(x))$ and $(g(x)\,\xi_r(x))$ are complete in $L^2((a, b), w)$.

(3) Each of the sequences so far mentioned in this section remains complete if we select only those members for which r ranges over the rational numbers in (a, b) or, indeed, only those for which r ranges over any dense subset of (a, b).

A modification like this last one is only to be expected, since the complete sets mentioned up to there had been considerably over-populated in view of the separability of L^2 space.

Moving further towards a proof of the Vitali criterion, we now state and prove:

LAURICELLA'S CRITERION   *Let (ψ_n) (the indexing set may be uncountable) be total in V (V as in §1.1.2) and let (ϕ_n) be total* with respect to (ψ_n), that is, each ψ_n is arbitrarily close to some linear combination $\Sigma a_n \phi_n$. Then (ϕ_n) is total in V.

This is proved by taking S_2 to be the closed linear span of $\{\psi_n\}$ and S_1 to be the closed linear span of $\{\phi_n\}$ and applying the 'chain of dense subsets principle' (§ 1.1.6).

Where V is a Hilbert space H, it is not required that either (ϕ_n) or (ψ_n) be orthogonal. If indeed (ϕ_n) is orthonormal it follows from Lauricella's criterion and Parseval's relation that a necessary and sufficient condition for it to be complete in H is that

$$\sum_n |(\phi_n, \psi_k)|^2 = \|\psi_k\|^2$$

hold for every ψ_k. When H is $L^2(a,b)$, for example, this relation reads

$$\sum_n \left| \int_a^b \phi_n \overline{\psi_k} \right|^2 = \int_a^b |\psi_k|^2.$$

If we now take (ψ_k) to be the complete sequence of step functions (χ_r), we obtain

VITALI'S COMPLETENESS CRITERION *Let*

$$\{\phi_n : n = 1, 2, ...\}$$

form an ON sequence in $L^2(a,b)$, a and b finite. Then (ϕ_n) is complete in $L^2(a,b)$ if and only if

$$\sum \left| \int_a^r \phi_n \right|^2 = r - a$$

for every $r \in (a,b)$.

COROLLARY *The Vitali criterion remains valid if r is only required to range over a dense subset of (a,b).*

Proof The 'only if' part is obvious. The 'if' part follows at once from modification (3) above.

An application of this corollary will be found in the theory of Walsh and Haar functions (§ 2.4).

The philosophy behind the Vitali criterion is basically this: The Parseval relation is a fundamental completeness criterion but cannot usually be used to demonstrate the completeness of a set since it would have to be verified for every member of the space H. Where this is not practicable, the Lauricella criterion is used to reduce the problem to that of verifying the Parseval relation only for every member of a complete set in H. In the L^2 case, the choice of a suitable class of step functions for this complete set yields the Vitali criterion. In turn, the Vitali criterion requires the summation of a series with parameter r for at least countably many values of r. The Dalzell criterion, to be given in the next section, further reduces the problem to the summation of a series of constants. However, it involves extra integrals and, whilst it has its uses, may be no easier to use in practice than the Vitali criterion.

Difficulties arising from the integrals in either of these criteria can sometimes be mitigated by introducing extra terms into the integrand; such modifications arise from the modifications to the step functions (χ_r) already mentioned. These, and other extensions to the case of L^2 spaces on regions in two and three dimensional Euclidean spaces, will be found in the problem sets. For an extension to $L^2(\mathbb{R}, \alpha)$ with discontinuous measure α, see § 2.3.1.

EXAMPLE   The trigonometrical set

$$\{e^{inx}/\sqrt{(2\pi)}: n = 0, \pm 1, \pm 2, \ldots\}$$

forms a complete sequence in $L^2(-\pi, \pi)$.

The orthonormality and the completeness of this very important sequence of functions are extremely well known. Here we shall show that the completeness can be easily demonstrated by Vitali's criterion, provided that two subsidiary formulae are known. These are Euler's formula

$$\sum_{n=1}^{\infty} \frac{(-1)^{n+1}}{n^2} \cos nx = \frac{\pi^2}{12} - \frac{x^2}{4} \quad (x \in [-\pi, \pi]),$$

and the special case

$$\sum_{n=1}^{\infty} n^{-2} = \pi^2/6.$$

According to Vitali's criterion we must show that

$$\sum_{-\infty}^{\infty} \left| \int_{-\pi}^{r} \frac{e^{inx}}{\sqrt{(2\pi)}} dx \right|^2 = \pi + r.$$

Now the term corresponding to $n = 0$ in this sum is

$$\left| \frac{1}{\sqrt{(2\pi)}} \int_{-\pi}^{r} dx \right|^2 = \frac{1}{2\pi} (\pi + r)^2,$$

and some elementary calculations yield

$$\frac{1}{\pi n^2} (1 - (-1)^n \cos rn)$$

for the general term. Vitali's criterion now reads

$$\frac{1}{2\pi} (\pi + r)^2 + \frac{2}{\pi} \sum_{n=1}^{\infty} \frac{1}{n^2} (1 - (-1)^n \cos rn) = \pi + r,$$

and this is easily verified by using the Euler formulae.

Problems

2.5 Show that the following form CON sequences in the L^2 space given:

(a) $\{e^{in\pi x} : n = 0, \pm 1, ...\} : L^2(-1, 1)$.

(b) $\{e^{2in\pi x} : n = 0, \pm 1, ...\} : L^2(0, 1)$.

(c) $\{[2/(b-a)]^{\frac{1}{2}} \sin [n\pi(x-a)/(b-a)] : n = 1, 2, ...\} : L^2(a, b)$.

(d) $\{[1/(b-a)]^{\frac{1}{2}}, [2/(b-a)]^{\frac{1}{2}} \cos [n\pi(x-a)/(b-a)] : n = 1, 2...\} : L^2(a, b)$.

(e) $\left\{ \frac{1}{\sqrt{(2\pi)}}, \frac{\cos nx}{\sqrt{\pi}}, \frac{\sin (n-\frac{1}{2})x}{\sqrt{\pi}} : n = 1, 2, ... \right\} : L^2(-\pi, \pi)$.

2.6 Use Vitali's criterion to demonstrate the completeness of the Legendre polynomials in $L^2(-1, 1)$.

2.7 Prove the following modified form of Vitali's criterion: Let (a, b) be a finite or infinite interval of $\mathbb{R}$, let g belong to $L^2((a, b), w)$, $g \neq \theta$, where w is a positive continuous weight function, and let (ϕ_n) be an ON sequence in $L^2((a, b), w)$. Then (ϕ_n) is complete in $L^2((a, b), w)$ (equivalently $(\phi_n \sqrt{w})$ is complete in $L^2(a, b)$) if and only if

$$\sum_n \left| \int_a^r \phi_n(x) g(x) w(x) dx \right|^2 = \int_a^r |g(x)|^2 w(x) dx$$

for every r in (a, b).

2.8 Show that the orthonormal associated Legendre functions

$$\left[\frac{2n+1}{2}\frac{(n-m)!}{(n+m)!}\right]^{\frac{1}{2}}P_n^m(x),$$

$n = m, m+1, \ldots$ (Appendix 2,6) are complete in $L^2(-1, 1)$, and hence obtain the formula

$$\sum_{n=m}^{\infty}\frac{2n+1}{2}\frac{(n-m)!}{(n+m)!}\left[\int_r^1 P_n^m(x)\,dx\right]^2 = 1-r$$

(Sansone (1959) p. 249).

2.9 Let S denote the surface of the sphere of unit radius in 3-space, and let (r, ϕ, θ) denote the usual spherical polar coordinates (ϕ the 'polar colatitude', θ the 'longitude'). Let $\{f_n\}$ be a set of functions which are orthonormal over S, i.e.

$$\int_S f_n \bar{f}_m = \delta_{nm}.$$

Obtain the following spherical form of the Vitali completeness criterion: The ON sequence (f_n) is complete in $L^2(S)$ if and only if

$$\sum_n \left[\int_0^l d\theta \int_0^p f_n(\phi, \theta)\sin\phi\,d\phi\right]^2 = l(1 - \cos p)$$

for every $l \in (0, 2\pi)$ and every $p \in (0, \pi)$ (Sansone (1959) p. 271).

2.10 Use the criterion of the previous problem to show that the set of 'spherical harmonics'

$$\left(\frac{2n+1}{4}\right)^{\frac{1}{2}}P_n(\cos\phi)$$

$$\left(\frac{2n+1}{2}\frac{(n-m)!}{(n+m)!}\right)^{\frac{1}{2}}P_n^m(\cos\phi)\cos m\theta$$

$$\left(\frac{2n+1}{2}\frac{(n-m)!}{(n+m)!}\right)^{\frac{1}{2}}P_n^m(\cos\phi)\sin m\theta \quad (n = 1, 2, \ldots),$$

is complete in $L^2(S)$. Hint: you will need the series of problem 2.7 (Sansone (1959) p. 271).

2.11 Formulate and prove a Vitali criterion applicable to $L^2(D)$, where D is the unit disc in the complex plane. Give an example of a CON sequence in $L^2(D)$.

2.3   The Dalzell completeness criterion

Let us return for a moment to the example at the end of the previous section and note that the required Vitali relation could have been verified using only the second of the Euler formulae. This is seen by formally integrating the relation to be verified with respect to r over $(-\pi, \pi)$, and obtaining

$$\frac{4\pi^2}{3} + 4 \sum_{n=1}^{\infty} \frac{1}{n^2} = 2\pi^2,$$

a formula which is easily seen to be true from the second Euler formula only. The method of Dalzell is to get rid of the dependence of the Vitali criterion on r by justifying such formal integrations.

DALZELL'S COMPLETENESS CRITERION (Dalzell, 1945a)
Let (a, b) be a finite interval and (ϕ_n) be an ON sequence in $L^2(a, b)$. Then (ϕ_n) is complete in $L^2(a, b)$ if and only if

$$\frac{2}{(b-a)^2} \sum_n \int_a^b \left| \int_a^r \phi_n(t)\, dt \right|^2 dr = 1.$$

Proof 'Only if' Vitali's criterion may be integrated between a and b and the order of integration and summation interchanged on the left-hand side, by the Levi theorem.

'If' This process of integration can be 'undone', as follows. Put

$$F(r) = a - r - \sum_n \left| \int_a^r \phi_n \right|^2$$

then we have by hypothesis

$$\int_a^b F(r)\, dr = 0.$$

Now Bessel's inequality shows that F is non-negative on (a, b), hence $F(r) = 0$, a.a. $r \in (a, b)$. But such a set of rs is dense in (a, b), so the proof is completed by appeal to the corollary to Vitali's criterion.

We have seen that the 'if' part of the Dalzell criterion provides a quick way of showing completeness of the trigonometrical functions, given the Euler formula $\Sigma n^{-2} = \pi^2/6$. From the 'only if' part it follows that this formula is actually equivalent to the completeness property of the trigonometrical functions.

Problems

2.12 Use Dalzell's method to demonstrate the completeness in $L^2(-1, 1)$ of the Legendre polynomials.

2.13 Obtain the following modified form of Dalzell's criterion: Let (a,b) be a finite or infinite interval of $\mathbb{R}$, (ϕ_n) an ON sequence in $L^2(a,b)$, $g \in L^2(a,b)$ $(g \neq \theta)$ and w a positive continuous weight function which is integrable over (a,b). Then (ϕ_n) is complete in $L^2(a,b)$ if and only if

$$\sum_n \int_a^b \left| \int_a^r \phi_n(x)\, g(x)\, dx \right|^2 w(r)\, dr = \int_a^b \left\{ \int_a^r |g(x)|^2\, dx \right\} w(r)\, dr.$$

A more general form of this criterion was discovered independently by Graves (1952).

Orthogonal sequences tend to fall into certain obvious categories, and whilst we do not wish to attempt a complete scheme of classification it will be worth noting some of these categories. For example there are the various sets of trigonometrical functions, including the exponential functions $\{e^{inx}\}$, constituting a category of elementary functions. Another category consists of polynomials, possibly 'weighted'. Again, there is a category containing higher transcendental functions typified by the example involving Bessel functions given below. Still other categories involve rational functions (§2.6.4), and discontinuous functions (§2.4). Non-orthogonal sequences fall into similar categories, particularly those formed by 'perturbing' ON sequences of a given category (see §§3.1.1, 3.3, 3.4).

EXAMPLE (completeness of the Fourier–Bessel functions) We shall use the modified Dalzell method (problem 2.13) to show that the set of ON Fourier–Bessel functions

$$\left\{ \frac{2^{\frac{1}{2}}}{|J_{\nu+1}(j_{n\nu})|} x^{\frac{1}{2}} J_\nu(j_{n\nu} x) : n = 1, 2, \dots \right\},$$

(see Appendix 2,9 for definitions) forms a complete sequence in $L^2(0, 1)$.

The inner integral on the left-hand side of the criterion to be verified is, except for the normalising factor,

$$\int_0^r x^{\frac{1}{2}} J_\nu(j_{n\nu} x) f(x) \, dx.$$

The choice $f(x) = x^{\nu+\frac{1}{2}}$ yields the well-known integral (Magnus *et al.* (1966) p. 86)

$$\int_0^r x^{\nu+1} J_\nu(j_{n\nu} x) \, dx = \frac{r^{\nu+1}}{j_{n\nu}} J_{\nu+1}(j_{n\nu} r).$$

To continue we must square this and evaluate

$$j_{n\nu}^{-2} \int_0^1 r^{2\nu+2} J_{\nu+1}^2(j_{n\nu} r) \, w(r) \, dr.$$

The choice $w(r) = r^{-2\nu-1}$ yields the known integral (Magnus *et al.* (1966) p. 88)

$$\int_0^1 r J_{\nu+1}^2(j_{n\nu} r) \, dr = -\frac{1}{2} \left\{ [J_{\nu+1}'(j_{n\nu})]^2 - \left[1 + \left(\frac{(\nu+1)}{j_{n\nu}} \right)^2 \right] J_{\nu+1}^2(j_{n\nu}) \right\}.$$

Now from the recurrence relation

$$\frac{\nu}{z} J_\nu(z) + J_\nu'(z) = J_{\nu-1}(z)$$

we have
$$\frac{-(\nu+1)}{j_{n\nu}} J_{\nu+1}(j_{n\nu}) = J_{\nu+1}'(j_{n\nu}),$$

so that the left-hand side of the Dalzell criterion reduces to $\Sigma j_{n\nu}^{-2}$. The right-hand side, with our particular choices for f and w, reduces by elementary integrations to $1/(4\nu+1)$. The series can be summed by the method of residues, for the function

$$\frac{1}{z^2} \frac{J_\nu'(z)}{J_\nu(z)} = \frac{\nu}{z^3} - \frac{1}{z} \frac{1}{2(\nu+1)} + O(z), \quad |z| \to 0$$

has a pole at the origin with residue $-1/2(\nu+1)$ and simple poles at $\pm j_{n\nu}$ with residues $j_{n\nu}^{-2}$. A sequence (C_k) of contours can be constructed so that

$$\lim_{k \to \infty} \frac{1}{2\pi i} \int_{C_k} \frac{1}{z^2} \frac{J_\nu'(z)}{J_\nu(z)} \, dz = 0$$

$$= \lim_{k \to \infty} 2 \sum_{n=1}^{k} j_{n\nu}^{-2} - 1/2(\nu+1).$$

That is
$$\sum_{n=1}^{\infty} j_{n\nu}^{-2} = 1/4(\nu+1),$$

and this verifies the required criterion.

It is of interest to point out that, if we had used the Vitali method and stopped short of the second integration process, we should have had to verify

$$\sum_{n=1}^{\infty} \left\{ \frac{1}{j_{n\nu}} \frac{J_{\nu+1}(j_{n\nu} r)}{J_{\nu+1}(j_{n\nu})} \right\}^2 = 1/4(\nu+1), \quad 0 \leqslant r \leqslant 1.$$

The fact that we have already proved completeness may be taken as a proof of this formula in the theory of the Bessel functions.

Problem

2.14 Let $\{\alpha_n\}$ be the positive zeros of
$$x J_\nu'(x) + h J_\nu(x), \quad h + \nu > 0.$$

Show that the set $\{\phi_n(x)\} = \{x^{\frac{1}{2}} J_\nu(\alpha_n x)\}$ is orthogonal over $(0, 1)$. Calculate the normalising factors for $\{\phi_n\}$ and show that the resulting ON set is complete in $L^2(0, 1)$ (Dalzell, 1945 b).

2.3.1 The Poisson–Charlier polynomials Let $\sigma(x)$ be a step function which is zero on $\mathbb{R}^-$ and takes the jumps σ_k at $x = k$, $k = 0$, 1, ..., with $\sigma_k > 0$ and $\sum_{n=0}^{\infty} \sigma_k = 1$. Thus, we may regard σ as a finite normalised measure on $\mathbb{R}$.

We shall consider the space $L^2(\mathbb{R}, \sigma)$. Note that the inner product of two members f_1 and f_2 of this space is given by

$$(f_1, f_2) = \int_{\mathbb{R}} f_1 \bar{f_2} \, d\sigma = \sum_{k=0}^{\infty} \sigma_k f_1(k) \bar{f_2}(k)$$

and that f is null if and only if $f(k) = 0$ $(k = 0, 1, ...)$.

We can develop a Vitali criterion for $L^2(\mathbb{R}, \sigma)$ as follows: first, *the step functions* $\{\chi_r\}$ (see p. 33) *form a complete sequence in* $L^2(\mathbb{R}, \sigma)$, for if f is any member of this space, and

$$\int_{\mathbb{R}} \chi_r f d\sigma = 0 \text{ for every } r,$$

then

$$\int_{-\infty}^{r} f d\sigma = 0 \text{ for every } r,$$

i.e.

$$\sum_{k=0}^{[r]} \sigma_k f(k) = 0 \text{ for every } r.$$

That is, f is null.

Suppose now that (ϕ_n) is an ON sequence in $L^2(\mathbb{R}, \phi)$. As before, the Parseval relation and Lauricella criterion yield

$$\sum_n \left| \int_{-\infty}^{r} \phi_n d\sigma \right|^2 = \int_{-\infty}^{r} d\sigma \text{ for every } r$$

as a necessary and sufficient condition for the completeness of (ϕ_n) in $L^2(\mathbb{R}, \sigma)$. But this condition is

$$\sum_n \left| \sum_{k=0}^{[r]} \sigma_k \phi_n(k) \right|^2 = \sigma(r) = \sum_{k=0}^{[r]} \sigma_k \text{ for every } r.$$

Thus we have the

VITALI CRITERION FOR $L^2(\mathbb{R}, \sigma)$ *Let σ be the step function defined above and (ϕ_n) an ON sequence in $L^2(\mathbb{R}, \sigma)$. Then in order that (ϕ_n) be complete in $L^2(\mathbb{R}, \sigma)$ it is necessary and sufficient that*

$$\sum_n \left| \sum_{k=0}^{r} \sigma_k \phi_n(k) \right|^2 = \sum_{k=0}^{r} \sigma_k \quad (r = 0, 1, \ldots).$$

EXAMPLE (The Poisson–Charlier polynomials) Let α be a step function as above, with the jumps $\alpha_k = e^{-a} a^k / k!$ at $k = 0, 1, \ldots$. The measure α is associated with the Poisson distribution in the theory of statistics. Charlier introduced a set of polynomials defined on the non-negative integers and orthogonal in $L^2(\mathbb{R}, \alpha)$ by the formula

$$p_n(r) = \frac{a^{n/2}}{(n!)^{\frac{1}{2}}} \sum_{k=0}^{n} (-1)^{n-k} \binom{n}{k} \binom{r}{k} k! \, a^{-k} \quad (n = 0, 1, \ldots),$$

(the sum actually terminates at the term for which $k = \min(n, r)$).

Now (p_n) is also complete in $L^2(\mathbb{R}, \alpha)$, a result apparently due to Szegö but not stated in such terms (see Schmidt (1933) and the literature cited there). We prove the orthogonality of (p_n), and as a consequence the completeness from the Vitali criterion just obtained, by the interesting and powerful method of generating functions.

We prove first the generating function relation

$$G(k, w) = e^{-w}(1 + w/a)^k = \sum_{n=0}^{\infty} \frac{a^{-n/2} p_n(k) w^n}{(n!)^{\frac{1}{2}}}, \quad |w| < a.$$

Now from the definition of p_n this sum is

$$\sum_{n=0}^{\infty} \sum_{m=0}^{n} \frac{(-1)^{n-m}}{n!} \binom{n}{m} \binom{k}{m} m! \, a^{-m} w^n$$

$$= \sum_{n=0}^{\infty} \sum_{m=0}^{\infty} \frac{(-1)^n}{(n+m)!} \binom{n+m}{m} \binom{k}{m} m! \, a^{-m} w^{n+m}$$

$$= \sum_{n=0}^{\infty} \sum_{m=0}^{\infty} \frac{(-1)^n}{n!} \binom{k}{m} \left(\frac{w}{a}\right)^m w^n$$

$$= e^{-w} \sum_{m=0}^{\infty} \binom{k}{m} \left(\frac{w}{a}\right)^m$$

$$= e^{-w}(1 + w/a)^k, \text{ as required.}$$

Thus we shall have

$$\sum_{k=0}^{\infty} \alpha_k \, G(k, u) \, G(k, v) = \sum_{k=0}^{\infty} \frac{e^{-a} a^k}{k!} \, e^{-u}(1 + a/u)^k \, e^{-v}(1 + a/v)^k$$

$$= e^{-a-u-v} \, e^{a(1+u/a)\,(1+v/a)}$$

$$= e^{uv/a},$$

and by comparing powers of uv on left- and right-hand sides, we find that

$$\sum_{k=0}^{\infty} \alpha_k \frac{p_n(k)}{a^{n/2}(n!)^{\frac{1}{2}}} \frac{p_m(k)}{a^{m/2}(m!)^{\frac{1}{2}}} = \frac{\delta_{nm}}{a^n n!}.$$

Thus, the orthogonality relation for the Poisson–Charlier polynomials is

$$\sum_{k=0}^{\infty} \alpha_k p_n(k) \, p_m(k) = \delta_{nm}.$$

But, notice that from the definition

$$(-1)^n \frac{(n\,!)^{\frac{1}{2}}}{a^{n/2}} \, p_n(k) = (-1)^k \frac{(k\,!)^{\frac{1}{2}}}{a^{k/2}} \, p_k(n),$$

so that an alternative orthogonality relation is

$$\sum_{k=0}^{\infty} p_k(n)\, p_k(m) = e^{-a} n\,!\, a^{-n} \, \delta_{nm}.$$

This last relation is useful when it comes to verifying the completeness of (p_n), for the Vitali criterion (above) requires us to verify

$$\sum_{m=0}^{\infty} \left\{ \sum_{j=0}^{r} p_m(j) \frac{a^j}{j\,!} \right\}^2 = e^a \sum_{j=0}^{r} \frac{a^j}{j\,!} \quad (r = 0, 1, \ldots),$$

the left-hand side of which can be written

$$\sum_{j=0}^{r} \left\{ \sum_{m=0}^{\infty} p_m^2(j) \right\} \left(\frac{a^j}{j\,!} \right)^2 + \sum_{\substack{i,\,j=0 \\ i<j}}^{r} \left\{ \sum_{m=0}^{\infty} p_m(j)\, p_m(i) \right\} \frac{a^{i+j}}{(i+j)\,!}.$$

Now the alternative orthogonality relation shows that the second sum is zero, and that the first is

$$e^{-a} \sum_{j=0}^{r} \frac{j\,!}{a^j} \left(\frac{a^j}{j\,!} \right)^2 = \sum_{j=0}^{r} \frac{a^j}{j\,!}.$$

This verifies the Vitali criterion, and we have shown that *The Poisson–Charlier polynomials form a CON sequence in* $L^2(\mathbb{R}, \alpha)$. This completes the example.

For further polynomials orthogonal over similar L^2 spaces, see Szegö (1939) and Carlitz (1960). Szegö proves a closure theorem for $L^p((a,b), \sigma)$, (a,b) a finite interval.

Problem

2.15 Obtain a Dalzell criterion for $L^2(\mathbb{R}, \sigma)$, and apply it to the Poisson–Charlier polynomials.

2.4   The functions of Rademacher, Walsh and Haar

We have seen that the set $\{\sqrt{2} \sin n\pi x : n = 1, 2, \ldots\}$ forms a CON sequence in $L^2(0, 1)$ (problem 2.5(c)). Let us put

$$s_n(x) = \operatorname{sgn}(\sin n\pi x),$$

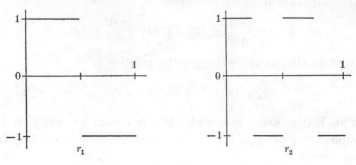

Fig. 1. The first two Rademacher functions.

$n = 1, 2, \ldots$: then (s_n) is a sequence of discontinuous functions which is not orthogonal over $(0, 1)$ but which is complete in $L^2(0, 1)$ (Harrington and Cell 1961). On the other hand that subset of $\{s_n\}$ consisting of the RADEMACHER FUNCTIONS, defined by
$$r_n(x) = \operatorname{sgn}(\sin 2^n \pi x) \quad (n = 1, 2, \ldots),$$
forms a sequence which is orthogonal over $(0, 1)$ (which the reader can easily verify) but is not complete in $L^2(0, 1)$. The failure of completeness arises from the fact that there are many functions orthogonal to all the Rademacher functions; for example, $\cos 2\pi x$ has this property and so does any L^2 function with the same symmetries.

The domain of the nth Rademacher function is divided into 2^{n-1} 'cycles' each of length $1/2^{n-1}$, over the first half of which r_n takes the value $+1$, and over the second, -1.

In order to 'complete' the Rademacher functions Walsh adjoined certain other combinations of Rademacher functions to them.

DEFINITION   Let r_n denote the nth Rademacher function; then we have the WALSH FUNCTIONS defined by
$$w_1(x) \equiv 1$$
$$w_{k+1}(x) = r_{n_1+1}(x)\, r_{n_2+1}(x) \ldots r_{n_N+1}(x), \quad k = 1, 2, \ldots,$$
where k is given its dyadic representation
$$k = 2^{n_1} + 2^{n_2} + \ldots + 2^{n_N}, \quad n_1 > \ldots > n_N \geqslant 0.$$

The cases $k = 2^{n_1}, n_1 = 0, 1, \ldots$, give the Rademacher functions themselves, so these are all contained in the Walsh system. An alternative definition in which the functions are 'sequency ordered', that is, the kth function has $k+1$ zero crossings on $(0, 1)$, has been developed for use in communication theory, where Walsh functions are used extensively (see Harmuth (1969) and Lackey and Meltzer (1971)). They are also important in the theory of probability.

Our present concern is with completeness properties, and our next theorem embodies the fundamental property of the Walsh functions.

THEOREM   *The Walsh system forms a CON sequence in* $L^2(0, 1)$

Proof (i) Orthonormality   A product of two Walsh functions will be of the form $(r_{m_1})^{a_1} (r_{m_2})^{a_2} \ldots (r_{m_p})^{a_p}$, where the ms are integers such that $m_1 > m_2 > \ldots m_p \geqslant 1$, and the as are all either 1 or 2. For the normality, all as will be 2, and since $r_n^2(x) = 1$ a.e. for every n it follows that the integral of the square of a Walsh function is 1. For orthogonality, we may delete from the product of two Walsh functions any squared Rademacher function and relabel the remaining subscripts $m_1, m_2, \ldots m_q$. Now

$$r_{m_2}(x) \ldots r_{m_q}(x)$$

is constant on each of the 2^{m_2} half-cycles of r_{m_2}. A typical half-cycle I_{m_2} is divided into $2^{m_1-m_2}$ half-cycles of r_{m_1}, in which r_{m_1} is alternately $+1$ and -1. Hence

$$\int_0^1 r_{m_1} \ldots r_{m_q} = \sum_{I_{m_2}} \int_{I_{m_2}} (\text{const.}) \, r_{m_1} = 0,$$

since
$$\int_{I_{m_2}} r_{m_1} = 0.$$

This proves the orthogonality.

(ii) Completeness   To satisfy the definition of completeness, let $f \in L^2(0, 1)$ and set

$$F(x) = \int_0^x f(t) \, dt.$$

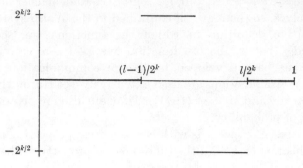

Fig. 2. A typical Haar function.

Then $F(0) = 0$.

Suppose $\displaystyle\int_0^1 f(x)\,w_{k+1}(x)\,dx = 0 \quad (k = 0, 1, \ldots)$,

then $\displaystyle\int_0^1 f(x)\,w_1(x)\,dx = 0 = F(1)$,

$$\int_0^1 f(x)\,w_2(x)\,dx = 0 = 2[F(\tfrac{1}{4}) + F(\tfrac{3}{4})] = 0,$$

$$\int_0^1 f(x)\,w_3(x)\,dx = 0 = 2[F(\tfrac{1}{4}) - F(\tfrac{3}{4})] = 0,$$

whence $F(\tfrac{1}{4}) = F(\tfrac{3}{4}) = 0$. Continuing in this way, use of w_4 yields

$$2[F(\tfrac{1}{8}) + F(\tfrac{3}{8}) + F(\tfrac{5}{8}) + F(\tfrac{7}{8})] = 0;$$

use of w_5, w_6, and w_7 yields the three similar equations in which two of the pluses are replaced with minuses, hence

$$F(\tfrac{1}{8}) = F(\tfrac{3}{8}) = F(\tfrac{5}{8}) = F(\tfrac{7}{8}) = 0.$$

In general we see that $F = 0$ on a dense subset of $(0, 1)$. By continuity $F \equiv 0$, $f(x) = 0$ a.e. and the proof is complete.

A more general completeness theorem of Rényi (Alexits (1961) p. 21) asserts that if $\{\phi_n(x)\}$ be any uniformly bounded set of measurable functions on $[a, b]$ which separates points of $[a, b]$ (i.e. given x_1, x_2 in $[a, b]$ then $x_1 \neq x_2$ implies $\phi_n(x_1) \neq \phi_n(x_2)$ for some n), then $\{\phi_1^{m_1}\phi_2^{m_2}\ldots\phi_n^{m_n}\}$ forms a complete sequence in

$L^2((a, b), \mu)$, where $m_k = 0, 1, \ldots; k = 1, 2, \ldots, n; n = 1, 2, \ldots$ and μ is a bounded, positive, monotone increasing measure with a non-negative derivative vanishing at most on a set of measure zero (see problem 2.16).

As well as the L^p theory there is a substantial literature on the pointwise convergence of series in Rademacher and Walsh functions. A similar system of discontinuous functions with an even more favourable convergence theory is that of Haar. The HAAR FUNCTIONS are defined by

$$h_1(x) \equiv 1$$

$$h_{2^k+l}(x) = \begin{cases} 2^{k/2}, x \in \left[\dfrac{l-1}{2^k}, \dfrac{l-\frac{1}{2}}{2^k}\right) \\[2ex] -2^{k/2}, x \in \left[\dfrac{l-\frac{1}{2}}{2^k}, \dfrac{l}{2^k}\right] \\[2ex] 0 \text{ otherwise,} \end{cases}$$

$l = 1, 2, \ldots, 2^k; k = 0, 1, \ldots$

The domain of a typical Haar function is divided into 2^k cycles of length $1/2^k$. For each cycle there is a Haar function with a Rademacher-type 'plus/minus' alternation of amplitude $2^{k/2}$ on that cycle, the function being 0 on all other cycles.

THEOREM *The Haar system forms a CON sequence in* $L^2(0, 1)$

Proof (*a*) Orthonormality The normality is obvious; the orthogonality follows from

(i) h_1 is orthogonal to all other Haar functions;

(ii) If $i \neq j$, $h_{2^k+i} h_{2^k+j} = 0$ a.e.;

(iii) If $n > m$, either $h_{2^n+1} h_{2^m+1} = 0$ a.e., or

$$\int_0^1 h_{2^n+i} h_{2^m+j} = \pm 2^{m/2} \int_0^1 h_{2^n+i} = 0.$$

(*b*) Completeness This can be proved by the direct method used to prove the completeness of the Walsh system; however, as an interesting contrast to this direct method we shall give a proof using the Vitali criterion.

The set of points $\{p/2^q\}$, where p is any odd positive integer and q any positive integer such that $p < 2^q$ is dense in $(0,1)$. Then if we can verify that

$$\sum_{r=1}^{\infty} \left\{ \int_0^{p/2^q} h_r(t)\,dt \right\}^2 = p/2^q,$$

Vitali's criterion (p. 35) will be satisfied. Put $r = 2^m + k$ and $\int_0^{p/2^q} h_r(t)\,dt = I(p,q,m,k)$. Now the non-zero values of h_r occur on a cycle of length 2^{-m}; hence $I(p,q,m,k) = 0$ if $2^{-q} \geqslant 2^{-m}$, i.e. $I(p,q,m,k) = 0$ unless $m \leqslant q-1$ (and hence $r \leqslant 2^q$). The equality to be verified now reads

$$\sum_{r=2}^{2^q} \{I(p,q,m,k)\}^2 = p2^{-q} - (p2^{-q})^2,$$

the term corresponding to $r = 1$ having been transposed to the right-hand side.

Now $p/2^q$ lies in exactly one cycle of h_r, so that given m there is exactly one k for which $I(p,q,m,k) \neq 0$. This k depends on p, q and m, and we shall denote the non-zero value of the integral by $I(p,q,m)$. Thus we must verify that*

$$\sum_{m=0}^{q-1} \{I(p,q,m)\}^2 = p2^{-q}(1 - p2^{-q}).$$

Our first task is to find an expression for $I(p,q,m)$. We give to p its dyadic representation:

$$p = \sum_{n=0}^{q-1} a_n 2^n, \quad a_n = 0 \text{ or } 1.$$

Put
$$S_{q-m-1} = \sum_{n=0}^{q-m-1} a_n 2^n,$$

so that $p = S_{q-1}$, and define $S_{-1} = 0$. Now we may write p as S_{q-m-2} plus terms with a common factor 2^{q-m-1}, thus

$$p = j2^{q-m-1} + S_{q-m-2}$$

---

* The author would like to thank Mr R. E. Abraham for supplying the elegant demonstration of this equality.

for some j. It is obviously convenient to have the range of integration written as a multiple of $1/2^{m+1}$, so that

$$\frac{p}{2^q} = \frac{p}{2^{q-m-1}} \frac{1}{2^{m+1}} = (j + S_{q-m-2}/2^{q-m-1})/2^{m+1}.$$

Now if $a_{q-m-1} = 0$, j is even and the integral over the range 0 to $j/2^{m+1}$ is zero. The integral over the remaining range is, taking account of the normalising factor, $2^{m/2}S_{q-m-2}/2^q$. If $a_{q-m-1} = 1$, j is odd and the integral over the range 0 to $j/2^{m+1}$ is $2^{m/2}/2^{m+1}$, the remainder giving $-2^{m/2}S_{q-m-2}/2^q$.

Since a_{q-m-1} is either 0 or 1, these results may be combined to give

$$
\begin{aligned}
I(p, q, m) &= 2^{m/2} a_{q-m-1}/2^{m+1} + (-1)^{a_{q-m-1}} 2^{m/2} S_{q-m-2}/2^q \\
&= 2^{m/2} [a_{q-m-1} 2^{q-m-1} + (-1)^{a_{q-m-1}} S_{q-m-2}]/2^q,
\end{aligned}
$$

and this completes the first part of the proof.

For the second part we shall denote $I(p, q, m)$ by I_m for brevity, and note two properties of the coefficients a_n in the dyadic representation of p: (i) $a_n^2 = a_n$, and (ii) $(-1)^{a_n} a_n = -a_n$. Now since

$$S_{q-m-1} = a_{q-m-1} 2^{q-m-1} + S_{q-m-2},$$

$$S_{q-m-1}^2 = a_{q-m-1} 2^{2q-2m-2} + a_{q-m-1} 2^{q-m} S_{q-m-2} + S_{q-m-2}^2.$$

Furthermore,

$$2^{q-m} I_m^2 = a_{q-m-1} 2^{2q-2m-2} - a_{q-m-1} 2^{q-m} S_{q-m-2} + S_{q-m-2}^2.$$

On adding these last two equations and rearranging, we get

$$S_{q-m-1}^2 - 2 S_{q-m-2}^2 + 2^{2q-m} I_m^2 = a_{q-m-1} 2^{2q-2m-1}.$$

On multiplying this by 2^m and summing on m from 0 to $q-1$, we get

$$\sum_{m=0}^{q-1} 2^m S_{q-m-1}^2 - \sum_{m=0}^{q-1} 2^{m+1} S_{q-m-2}^2 + \sum_{m=0}^{q-1} 2^{2q} I_m^2 = \sum_{m=0}^{q-1} a_{q-m-1} 2^{2q-m-1}.$$

Thus

$$S_{q-1}^2 - 2^q S_{-1}^2 + 2^{2q} \sum_{m=0}^{q-1} I_m^2 = 2^q \sum_{m=0}^{q-1} a_{q-m-1} 2^{q-m-1} = 2^q S_{q-1}$$

or
$$p^2 + 2^{2q} \sum_{m=0}^{q-1} I_m^2 = 2^q p.$$

Finally,
$$\sum_{m=0}^{q-1} I^2(p, q, m) = p 2^{-q}(1 - p 2^{-q})$$

as required.

Not only is this equation sufficient for the completeness of the Haar system, it is also necessary, as the Vitali criterion shows.

Problems

2.16 Use Rényi's theorem (p. 48) to show that the set

$$\{\cos^n x, \sin nx \cdot \cos^n x : n = 1, 2, \ldots\}$$

forms a complete sequence in $L^2((0, 2\pi), \mu)$. Deduce the completeness of the trigonometrical functions.

2.17 Show that

(i) $r_k(x) = 2^{-\frac{1}{2}(k-1)} \sum_{j=1}^{2^{k-1}} h_{2^{k-1}+j}(x)$ $(k = 1, 2, \ldots)$

(ii) $w_{2^k+l}(x) = r_{k+1}(x) w_l(x)$ $(l = 1, 2, \ldots, 2^k; k = 0, 1, \ldots)$

(iii) $w_{2^k+l}(x) = 2^{-k/2} \sum_{j=1}^{2^k} w_l(x) h_{2^k+j}(x)$
$$(l = 1, 2, \ldots, 2^k; k = 0, 1, \ldots).$$

2.18 Show that both the Walsh system and the Haar system are total in $L^p(0, 1,)$, $1 \leqslant p < \infty$. (Actually both are bases for $L^p(0, 1)$, see Singer (1970) p. 13 and p. 405.)

2.19 Show that for the Walsh system to be complete it is necessary and sufficient that

$$\sum_{m=1}^{q} 2^{m-1} \{J(p, q, m)\}^2 = p 2^{-q} (1 - p 2^{-q}),$$

where
$$J(p, q, m) = \int_0^{p/2^q} r_m(t) \, dt,$$

r_m being the mth Rademacher function (Higgins *et al.* 1975).

2.20 Let $g(x) \in L^2(0, 1)$, $g \neq \theta$ and be defined everywhere on $(0, 1)$ except possibly on a set of measure zero. Let $(g\chi_r)$ be the

complete sequence obtained by letting r take the values
$1/2$, $1/4$, $3/4$, $1/8$, $3/8$, $5/8$, $7/8$, ... (see p. 48). Find the
first few members of the CON sequence for $L^2(0, 1)$ obtained
by orthogonalising this set over $(0, 1)$. What happens if g
itself is expanded in the resulting CON sequence? If $g \equiv 1$,
show that the CON sequence is neither the Walsh nor the
Haar system, but that its members do have similar intervals
of constancy.

2.5 CON sequences and the reproducing kernel

Certain Hilbert spaces possess a 'reproducing kernel', which
gives them an even stronger structure than they would other-
wise have. This kernel can be calculated from a CON sequence
for the space. L^2 spaces do not normally possess a reproducing
kernel, but we shall shortly meet certain subspaces of L^2 which
do; consequently, we shall give the definition and a few important
facts in this section.

DEFINITION   A Hilbert space H of functions defined everywhere on a set X is called a 'reproducing kernel Hilbert space'
('r.k. space' for short) if there exists a kernel $k(t, x)$ defined on
$X \times X$, such that

  (i) $k(t, x) \in H$ $(x \in X)$,

  (ii) $f(x) = (f, k(., x))$ $(f \in H)$, the *reproducing equation*.

Note the necessity of using a dot in the notation; in the L^2 case,
this can be thought of as the dummy of integration. Before
pointing out some of the properties of r.k. spaces we give a
theorem and corollary which tell us when a Hilbert space will
have a reproducing kernel.

THEOREM   *A Hilbert function space is an r.k. space if and
only if the evaluation functional is bounded on H, i.e.*

$$|f(x)| \leqslant M_x \|f\| (f \in H).$$

Proof   (i) Let H be an r.k. space. Then

$$|f(x)| = |(f, k(., x))| \leqslant \|f\| \, \|k(., x)\|$$

so the evaluation functional is bounded.

(ii) Let the evaluation functional be bounded. Then by the Riesz–Fréchet representation theorem (Appendix 1,4), for each $x \in X$ there is a $g_x \in H$ such that $f(x) = (f, g_x(.))$. Then $g_x(t)$ has the required reproducing property.

COROLLARY* *If each $f \in H$ is continuous, then H is an r.k. space.*

Proof Define a set of linear functionals $\{T_\epsilon\}$ on H by $T_\epsilon f = (f, h_\epsilon)$ where $\int h_\epsilon = 1$ and $h_\epsilon(t)$ has support on $\{t : |x - t| < \epsilon\}$. Since $|(f, h_\epsilon)| \leqslant \|f\| \|h_\epsilon\|$ we find that $\{T_\epsilon\}$ is a set of bounded linear functionals on H. Furthermore, $\lim_{\epsilon \to 0} T_\epsilon f = f(x)$, since f is continuous, hence there exists a constant M_f such that

$$\sup_\epsilon |T_\epsilon f| < M_f \quad (f \in H).$$

The uniform boundedness principle applies and there exists M such that $\|T_\epsilon\| < M$ for every ϵ, hence

$$|T_\epsilon f| < \|f\| M.$$

We let $\epsilon \to 0$, then

$$|f(x)| \leqslant \|f\| M,$$

and by the previous theorem H is an r.k. space.

The converse of this corollary is not true; indeed, Lehto (1952) has given an example of an r.k. space which contains discontinuous functions.

Some useful properties of r.k. Hilbert spaces are listed; the reader will find that they follow readily from the definitions. For these and other properties, see Meschkowski (1962).

(a) The kernel $k(t, x)$ is unique.

(b) $k(t, x) = \overline{k(x, t)}$.

(c) $k(t, x) = (k(., t), k(., x))$; in particular $k(x, x) = \|k(., x)\|^2$.

(d) The weak and, hence, the strong convergence of a sequence (f_n) to f in an r.k. Hilbert space H implies the pointwise

---

* This result is due to Professor J. I. Richards, who communicated it to the author whilst supervising his doctoral research.

convergence of $f_n(x)$ to $f(x)$ $(x \in X)$. The convergence is uniform over any subset of X for which $k(x, x)$ is bounded.

(e) Let (ϕ_n) be a CON sequence in H. Then

$$k(t, x) = \Sigma \phi_n(t) \, \overline{\phi_n(x)} \quad (x, t \in X).$$

It is usually difficult to relate two different modes of convergence, therefore (d) is a remarkable feature of r.k. Hilbert spaces.

2.6 The method of isometric transformation

2.6.1 Description of the method Let H and K be two Hilbert spaces and $u: H \to K$ be an isometric isomorphism of H onto K; thus u preserves norms and hence inner products (see problem 2.21). It is evident then that u preserves orthonormality: *the image of an ON sequence in H is an ON sequence in K*. The mapping u also preserves completeness, whether the sequence is orthogonal or not, since every $k \in K$ is of the form $u(h)$ for some $h \in H$ and so if (ϕ_n) is complete in H, given $\epsilon > 0$, there exists a finite sequence of scalars (a_n) such that (using totality)

$$\|h - \Sigma a_n \phi_n\| = \|k - \Sigma a_n u(\phi_n)\| < \epsilon.$$

We could have started the argument with K and used u^{-1}, thus we have the important result that, *if u is an isometric isomorphism between two Hilbert spaces H and K, then a sequence in H and its image by u in K both possess or both fail to possess any or all of the properties*: completeness, orthogonality, normality (see also problem 2.21).

This is the essence of the isometric transform method, for, although H and K are, from the abstract point of view, entirely equivalent under u, this equivalence can be turned to good advantage in practice. For example, it may be easier to demonstrate completeness in H of the image sequence by a judiciously chosen u which maps H onto itself, rather than that of the original sequence. One can often (but not always—see problem 2.25) generate a new CON sequence for H in this way. A powerful application of the method is to the case where u maps $L^2(\mathbb{R})$ to itself, carrying a subspace of functions which vanish over some subset of $\mathbb{R}$ into a subspace which has some special property. In § 2.6.3 a class of interpolating functions will be generated in this way, and in § 2.6.4 a class of rational functions.

Problem

2.21 Show that an isometric isomorphism preserves (*a*) inner products, and (*b*) bases of Hilbert space.

An isometric isomorphism is not the most general type of mapping which preserves bases, see the Lemma on p. 75.

2.6.2 Examples of isometric transformations Before proceeding to illustrate the method, we shall give a few examples, some of which will be found useful later.

(i) The Fourier transform on $L^2(\mathbb{R})$. Plancherel's theory shows that if $f \in L^2(\mathbb{R})$ then the Fourier transform

$$(Ff)(x) = \underset{A\to\infty}{\text{l.i.m.}} \frac{1}{\sqrt{(2\pi)}} \int_{-A}^{A} e^{-ixt} f(t)\, dt,$$

where l.i.m. stands for 'limit in the mean', and indicates a limit in the sense of L^2 norm, is in $L^2(\mathbb{R})$. The Parseval relation $(f, g) = (Ff, Fg)$ holds for every f, $g \in L^2(\mathbb{R})$ and F is a unitary operator on $L^2(\mathbb{R})$. Under mild restrictions on f the integral defining Ff exists both in the sense of l.i.m. and in the sense of the ordinary limit, the two values being equal (see e.g. Titchmarsh (1937) p. 83).

(ii) The Fourier sine transform

$$(F_s f)(x) = \underset{A\to\infty}{\text{l.i.m.}} \frac{1}{\sqrt{(2\pi)}} \int_{0}^{A} \sin xt f(t)\, dt,$$

the Fourier cosine transform

$$(F_c f)(x) = \underset{A\to\infty}{\text{l.i.m.}} \frac{1}{\sqrt{(2\pi)}} \int_{0}^{A} \cos xt f(t)\, dt,$$

and the Hankel transform

$$(H_\nu f)(x) = \underset{A\to\infty}{\text{l.i.m.}} \int_{0}^{A} (xt)^{\frac{1}{2}} J_\nu(xt) f(t)\, dt, \quad \nu > -1,$$

are all examples of unitary operators on $L^2(\mathbb{R}^+)$.

The examples (i) and (ii) are treated in detail by Titchmarsh (1937) together with more general classes of unitary integral operators.

(iii) Let the function $\phi(t)$ have a positive derivative everywhere on an interval (c, d), with $\phi(c) = a$, and $\phi(d) = b$, and $\psi(t)$ be real and measurable. The mapping which takes $f(t)$ to $f(\phi(t)) [\phi'(t)]^{\frac{1}{2}} e^{i\psi(t)}$ provides an isometry of $L^2(a, b)$ into $L^2(c, d)$, as may easily be verified. When $\phi'(t)$ is everywhere negative, the mapping $f(t) \to f(\phi(t)) |\phi'(t)|^{\frac{1}{2}} e^{i\psi(t)}$ takes $L^2(a, b)$ to $L^2(d, c)$; for example with $\phi(t) = e^t$ the first maps $L^2(\mathbb{R}^+)$ to $L^2(\mathbb{R})$, and with $\phi(t) = -\log t$ the second maps $L^2(\mathbb{R}^+)$ to $L^2(0, 1)$.

O. Szász used several such transformations to study completeness properties of sequences (x^{λ_n}) (see p. 97).

2.6.3 The functions of the cardinal series

We shall use the isometric transform method to discuss the functions appearing in the series

$$\sum_{-\infty}^{\infty} a_n \frac{\sin \pi(x-n)}{\pi(x-n)},$$

which is called a *cardinal series*; we shall also point out some properties of this series. The method is due to Hardy (1941).

We have seen that $\{(2\pi)^{-\frac{1}{2}} e^{inx} : n = 0, 1, \ldots\}$, forms a CON sequence in $L^2(-\pi, \pi)$. Consider the set $\{\phi_n : n = 0, 1, \ldots\} \subset L^2(\mathbb{R})$ given by

$$\phi_n(x) = \begin{cases} (2\pi)^{-\frac{1}{2}} e^{inx}, & |x| < \pi, \\ 0, & |x| > \pi. \end{cases}$$

This forms an ON sequence over $\mathbb{R}$, which is not of course complete in $L^2(\mathbb{R})$. Now

$$\hat{\phi}_n(x) = F\phi_n(x) = \frac{1}{2\pi} \int_{-\pi}^{\pi} e^{int-ixt} \, dt$$

$$= \frac{\sin \pi(x-n)}{\pi(x-n)},$$

and $\{\hat{\phi}_n : n = 0, 1, \ldots\}$, is ON over $\mathbb{R}$ by the isometric property of F. It does not form a complete sequence in $L^2(\mathbb{R})$ since (ϕ_n) does not. Can we, then, identify that subspace F^π of $L^2(\mathbb{R})$ which is spanned by $(\hat{\phi}_n)$? To do this, we attempt to verify the definition of totality by taking $g \in L^2(\mathbb{R})$ and assuming that $(g, F\phi_n) = 0$

$(n = 0, \pm 1, ...)$. But this is equivalent to $(F^{-1}g, \phi_n) = 0$
$(n = 0, \pm 1, ...)$, where the definition of ϕ_n shows that the inner
product is taken over $[-\pi, \pi]$. Thus the completeness of (ϕ_n)
when restricted to $L^2(-\pi, \pi)$ shows that $F^{-1}g$ is null on $[-\pi, \pi]$.
It will be null on the whole of $\mathbb{R}$ if, and only if, g belongs to that
subclass of $L^2(\mathbb{R})$ whose inverse Fourier transform is null on the
complement of $[-\pi, \pi]$. This provides us with the subspace F^π
that we are seeking. The functions g of F^π may thus be written

$$g(x) = \frac{1}{\sqrt{(2\pi)}} \int_{-\pi}^{\pi} f(t) e^{-ixt} dt. \tag{1}$$

Hardy called these the Paley–Wiener functions, in view of the
well-known characterisation of such functions as entire functions
of exponential type whose restriction to the real axis is in
$L^2(\mathbb{R})$ (Paley and Wiener, 1934). F^π is an r.k. Hilbert space (each
$g \in F^\pi$ is continuous) in which

$$\left\{ \frac{\sin \pi(x-n)}{\pi(x-n)} : n = 0, \pm 1, ... \right\},$$

forms a CON sequence (see problem 2.22 for the reproducing
kernel). The Fourier series in this CON sequence for any $g \in F^\pi$
is a cardinal series; from the reproducing kernel theory we find
that the series converges pointwise and uniformly over $\mathbb{R}$, and
therefore reduces to a_n when x is an integer n. That is, $a_n = g(n)$
so that the formula

$$g(x) = \sum_{-\infty}^{\infty} g(n) \frac{\sin \pi(x-n)}{\pi(x-n)} \tag{2}$$

holds uniformly on $\mathbb{R}$ for each $g \in F^\pi$.

The above remarks have recently seen important application
in the theory of communication of information: here g is a signal
(one writes t instead of x to denote time) and provided that the
signal is 'band-limited', or 'slowly oscillating', i.e. contains no
frequency outside the 'band' $[-\pi, \pi]$ as in the representation
(1) then it can be completely constructed (theoretically at least)
by the formula (2) using only its values at a set of equally
spaced instants of time.

The cardinal series and its interpolating property were introduced by E. T. Whittaker (1915) who also studied the slowly oscillating property of its sum.

Problems

2.22 Show that the reproducing kernel for F^π is

$$\frac{\sin \pi(x-y)}{\pi(x-y)}.$$

2.23 Show that there is an interpolation series involving Bessel functions analogous to the cardinal series by using the fact that the Hankel transform (p. 56) is a unitary operator on $L^2(0, \infty)$, and by using the completeness of the Fourier–Bessel functions in $L^2(0, 1)$. Show also that the Hilbert space B_ν spanned by the functions of this series is that subspace of $L^2(0, \infty)$ consisting of functions whose Hankel transform of order ν is null outside $(0, 1)$ (Higgins, 1972).

2.24 Show that of the two ON sets of functions associated with the Bessel–Neumann series, namely,

$$\{2^{-1}(2n+1)^{\frac{1}{2}}(-1)^n x^{-\frac{1}{2}} J_{n+\frac{1}{2}}(x): n = 0, 1, ..., x \in \mathbb{R}\}$$

and

$$\{[2(\nu+2n+1)]^{\frac{1}{2}} x^{-\frac{1}{2}} J_{\nu+2n+1}(x): n = 0, 1, ..., x \in \mathbb{R}^+\},$$

the first forms a CON sequence in F^π, the second in B_ν. Hint: use the Hankel and the Fourier transform (Higgins, 1972).

2.6.4 The Laguerre polynomials and their Fourier Transforms   Here we shall discuss the completeness of the generalised Laguerre polynomials, and use the isometric transform method to generate a class of rational functions which forms a CON sequence in $L^2(\mathbb{R})$.

We have already observed that the completeness theorem for polynomials applies to the Laguerre polynomials in $L^2(\mathbb{R}^+, e^{-x})$. It can be easily verified that the theorem also applies to the polynomials obtained by orthonormalising the set of powers $\{x^n: n = 0, 1, ...\}$ with respect to $L^2(\mathbb{R}^+, x^\alpha e^{-x})$, so that a CON sequence in $L^2(\mathbb{R}^+)$ is formed from the set $\{l_n(x, \alpha): n = 0, 1, ...\}$,

where
$$l_n(x, \alpha) = \left[\frac{n!}{\Gamma(1+n+\alpha)} \right]^{\frac{1}{2}} x^{\alpha/2} e^{-x/2} L_n^{(\alpha)}(x).$$

Here, $L_n^{(\alpha)}(x)$ is the nth generalised Laguerre polynomial (Appendix 2,7). The completeness can be made to depend (Tricomi, 1955) on the formula

$$r^{2(\alpha+1)} \sum_{n=1}^{\infty} \frac{n!}{\Gamma(2+n+\alpha)(1+n+\alpha)} \{L_n^{(\alpha+1)}(r)\}^2 = \int_0^r x^\alpha e^x dx,$$
(3)

known from the theory of the 'incomplete gamma function'. This connection emerges from using the modified Vitali criterion (p. 37) with $w(x) = e^{-x} x^\alpha$ and the choice $g(x) = e^x$. Then the equality to be verified is

$$\sum_{n=1}^{\infty} \frac{n!}{\Gamma(1+n+\alpha)} \left\{ \int_0^r x^\alpha L_n^{(\alpha)}(x) dx \right\}^2 = \int_0^r e^x x^\alpha dx.$$

Now if we substitute
$$\int_0^r x^\alpha L_n^{(\alpha)}(x) dx = \frac{r^{\alpha+1}}{1+n+\alpha} L_n^{(\alpha+1)}(r)$$

(Magnus *et al.* (1966) p. 241), the equality to be verified is just the formula (3). Thus the completeness of the Laguerre polynomials and the formula (3) are consequences of each other. A modified Dalzell method can also be used, but it too involves difficult integrals (Dalzell, 1945a).

At this point we shall introduce a further parameter into the definition of the Laguerre polynomials. By a simple change of variable in the integral expressing orthonormality of $(L_n^{(\alpha)})$ we can show that a more general CON sequence in $L^2(\mathbb{R}^+)$ is formed from the set $\{l_n(x, \alpha, a): n = 0, 1, \ldots\}$, where

$$l_n(x, \alpha, a) = \left[\frac{a^{1+\alpha} n!}{\Gamma(1+n+\alpha)} \right]^{\frac{1}{2}} e^{-ax/2} x^{\alpha/2} L_n^{(\alpha)}(ax).$$

Next we calculate the Fourier transforms (see p. 56, (i) for the definition) of those functions which are equal to zero for negative arguments and equal to $l_n(x, \alpha, a)$ for positive arguments. We shall need

$$I = \frac{1}{\sqrt{(2\pi)}} \int_0^\infty (ax)^m x^{\alpha/2} e^{-ax/2} e^{-ixy} dx,$$

since the polynomial representation (Appendix 2,7) for $L_n^{(\alpha)}$ is to be used. Put $x(iy + a/2) = t$, then

$$I = \frac{a^m}{\sqrt{(2\pi)}} \int_C t^{m+\alpha/2} e^{-t}(iy + a/2)^{-1-m-\alpha/2} dt,$$

where C is a contour in the complex t-plane consisting of the whole ray $\arg t = 2y/a$. By considerations of regularity we may take C to be the ray $\arg t = 0$, and using the definition of the gamma function (Appendix 2,1) we find that

$$I = \frac{a^m \Gamma(1 + m + \alpha/2)}{\sqrt{(2\pi)}\,(iy + a/2)^{1+m+\alpha/2}}.$$

Hence

$$Fl_n(., \alpha, a)\,(y)$$

$$= \left[\frac{a^{1+\alpha}\,n\,!}{2\pi\Gamma(1+n+\alpha)} \right]^{\frac{1}{2}} \sum_{m=0}^{n} (-1)^m \binom{n+\alpha}{n-m} \frac{a^m}{m\,!} \frac{\Gamma(1+m+\alpha/2)}{(iy+a/2)^{1+m+\alpha/2}}.$$

$$(4)$$

First consider the special case $\alpha = 0$, $a = 2$. We have

$$Fl_n(., 0, 2)\,(y) = (\pi)^{-\frac{1}{2}} \sum_{m=0}^{n} \binom{n}{m} \frac{(-2)^m}{(1+iy)^{1+m}}$$

$$= \frac{(\pi)^{-\frac{1}{2}}}{(1+iy)} \left[1 - \frac{2}{(1+iy)} \right]^n$$

$$= (\pi)^{-\frac{1}{2}} \frac{(iy-1)^n}{(1+iy)^{n+1}}$$

$$= \rho_n(y), \text{ say}.$$

By an argument similar to that of the previous section, the set $\{\rho_n\}$ forms a CON sequence in that subspace F^+ of $L^2(\mathbb{R})$ whose inverse Fourier transforms vanish on the negative real axis; such functions are important in time series analysis (Wiener, 1949).

For general a and α the formulae in Appendix 2 for the binomial coefficient, the Pochhammer symbol and the hypergeometric function $_2F_1$ may be used, together with

$$(n-m)\,! = (-1)^m n\,!/(-n)_n,$$

to put (4) into the form

$$\left[\frac{a^{1+\alpha} n! \Gamma(1+n+\alpha)}{n! 2\pi}\right]^{\frac{1}{2}} \frac{1}{n!(iy+a/2)^{1+\alpha}}$$

$$\times \left[\sum_{m=0}^{n-1} \frac{(-n)_m a^m \Gamma(1+\alpha/2)(1+\alpha/2)_m}{m! \Gamma(1+\alpha)(1+\alpha)_m (iy+a/2)^m} + \frac{(-1)^n a^n (1+n+\alpha/2)}{\Gamma(1+n+\alpha)(iy+a/2)^n}\right]$$

$$= \left[\frac{a^{1+\alpha} \Gamma(1+n+\alpha)}{n! 2\pi}\right]^{\frac{1}{2}} \frac{\Gamma(1+\alpha/2)}{\Gamma(1+\alpha)} \frac{1}{(iy+a/2)^{1+\alpha}}$$

$$\times {}_2F_1\left(-n, 1+\alpha/2, 1+\alpha, \frac{2a}{2iy+a}\right).$$

This formula provides a class of CON sequences in F^+.

We now turn to the completeness of the rational functions $\{\rho_n : n = 0, 1, \ldots\}$ defined above. We have

$$\overline{\rho_{-n}(x)} = -\frac{(ix-1)^{n-1}}{\sqrt{\pi}(ix+1)^n} = -\rho_{n-1}(x),$$

and since $\rho_n(x) = Fl_n(., 0, 2)(x)$ we have

$$\rho_{-n}(x) = -\overline{\rho_{n-1}(x)} = -\overline{Fl_{n-1}(., 0, 2)(x)}$$

$$= \frac{\sqrt{2}}{\sqrt{(2\pi)}} \int_0^\infty e^{-t} L_{n-1}(2t) e^{ixt} dt = \frac{1}{\sqrt{\pi}} \int_{-\infty}^0 e^t L_{n-1}(-2t) e^{-ixt} dt.$$

This is the Fourier transform of the function equal to $\sqrt{2} e^t L_n(-2t)$ for negative arguments and equal to zero for positive arguments.

LEMMA *The set $\{\sqrt{2} e^t L_n(-2t) : n = 0, 1, \ldots\}$ forms a* CON *sequence in $L^2(\mathbb{R}^-)$.*

Proof (i) Orthonormality This is obtained from that of $(L_n(2t))$ over $\mathbb{R}^+$:

$$2\int_0^\infty e^{-2t} L_m(2t) L_n(2t) dt = \delta_{nm} = -2\int_0^{-\infty} e^{2t} L_m(-2t) L_n(-2t) dt.$$

(ii) Completeness Now the set $\{e^{-t} L_n(2t)\}$ forms a complete sequence in $L^2(\mathbb{R}^+)$, therefore if f is even and $f \in L^2(\mathbb{R})$,

$$-\int_0^{-\infty} f(t) e^t L_n(-2t) dt = \int_0^\infty f(t) e^{-t} L_n(2t) dt = 0 \quad (n = 0, 1, \ldots)$$

implies that f is null on $\mathbb{R}^+$, and hence null on $\mathbb{R}^-$. But every member of $L^2(\mathbb{R}^-)$ is the restriction to $\mathbb{R}^-$ of an even member of $L^2(\mathbb{R})$.

By Fourier transformation we now see that $\{\rho_{-n}(x):n = 1, 2, ...\}$ forms a CON sequence in F^-, the class of L^2 functions whose Fourier transforms vanish on $\mathbb{R}^+$. But it is clear that

$$L^2(\mathbb{R}) = F^+ \oplus F^-,$$

so that $\{\rho_n(x):n = 0, \pm 1, ...\}$ forms a CON sequence in $L^2(\mathbb{R})$. We have shown that the set

$$\left\{\frac{1}{\sqrt{\pi}}\frac{(ix-1)^n}{(ix+1)^{n+1}} : n = 0, \pm 1, ...\right\}$$

forms a CON sequence in $L^2(\mathbb{R})$.

Problems

2.25 Show that, apart from a factor i^n, the weighted Hermite polynomials

$$(\pi^{\frac{1}{2}}n!\,2^n)^{-\frac{1}{2}}e^{-x^2/2}H_n(x), \quad n = 0, 1, ...,$$

(see Appendix 2,8 for the definition of H_n), which form a CON sequence in $L^2(\mathbb{R})$ (problem 2.1), are their own Fourier transforms.

2.26 Show that $(\rho_n(x))$ is complete by the Dalzell criterion.

2.27 Let $\{\psi_n\}$ denote the weighted Laguerre polynomials on $\mathbb{R}^+$. Use the generating function relation

$$(1-t)^{-1}\exp\left\{-\frac{1}{2}\left(\frac{1+t}{1-t}\right)x\right\} = \sum_{n=0}^{\infty} t^n \phi_n(x)$$

to demonstrate the orthogonality of (ϕ_n). Use it also to show completeness in $L^2(\mathbb{R}^+)$, by the following steps:

(i) The series converges to the generating function in $L^2(\mathbb{R}^+)$ norm, for $|t| < 1$.

(ii) The set $\{\phi_n\}$ is total in the class of functions of the form $e^{-\alpha x}$, $x \in \mathbb{R}^+$, for any $\alpha \in \mathbb{R}^+$.

(iii) Consider the isometric transformation $L^2(\mathbb{R}^+) \to L^2(0, 1)$ induced by $f(x) \to i\, f(-\log x)/\sqrt{x}$ (see (iii) of p. 57); use the completeness of the powers in $L^2(0,1)$, then reverse the transformation.

This method was suggested by J. von Neumann.

2.28 Show, by calculating the inner products directly, that

$$\left\{ \frac{(ix-1)^{n+\rho}}{(ix+1)^{n+1+\rho}} : n = 0, \pm 1, \dots, \rho \text{ real} \right\},$$

forms an orthogonal sequence over $\mathbb{R}$. Show that this sequence is also complete in $L^2(\mathbb{R})$.

Show also that

$$\left\{ \left(\frac{a-ix}{a+ix} \right)^n : n = 0, \pm 1, \dots \right\}$$

forms a CON sequence in $L^2\left(\mathbb{R}, \dfrac{a}{\pi(a^2+x^2)} \right)$.

2.7 CON sequences of complex functions

In this section we shall investigate some CON sequences for L^2 spaces of functions regular in regions of the complex plane. Accordingly, let R denote a region (an open connected set) of the complex plane and let $L_r^2(R)$ denote the Hilbert space of all functions which are regular in R and such that

$$\|f\| = \iint_R |f(z)|^2 \, dx \, dy < \infty.$$

Note carefully that $L_r^2(R)$ is not the same as $L^2(R)$, the full L^2 space over R (see problem 2.30).

We are going to show that a conformal mapping of the underlying region R onto another region R' induces an isometric isomorphism of $L_r^2(R)$ onto $L_r^2(R')$. Readers wishing to strengthen their knowledge of conformal mapping should consult a good text, such as Nehari (1952).

Given a region R of the complex plane it is natural to seek a CON sequence of polynomials for $L_r^2(R)$. To this end, we can orthonormalise the sequence of powers (z^n) over R to obtain an ON set of polynomials. A pleasant feature of (z^n) is that it is already orthogonal over D, the unit disc. It is also complete in $L_r^2(D)$ (see the theorem to follow) these facts being particularly convenient since D is the canonical region for the Riemann

mapping theorem. For general R the ON sequence of polynomials will only be complete if R is reasonably shaped (see
theorem p. 68).

THEOREM $\{\sqrt{(n/\pi)}\, z^{n-1}: n = 1, 2, \ldots\}$ *forms a CON sequence*
in $L_r^2(D)$.

Proof (i) Orthonormality Now if $n \neq m$,

$$(z^n, z^m) = \iint_D z^n \bar{z}^m \, dx \, dy.$$

We can put $z = r e^{i\theta}$ and pass to polar coordinates to obtain the
value zero for this integral. Alternatively, the very useful
complex form of Green's theorem can be used, namely

$$\iint_R f(z)\,\overline{g'(z)}\, dx\, dy = \frac{1}{2i} \int_C f(z)\,\overline{g(z)}\, dz,$$

where C is the boundary of R (not necessarily simply connected),
and f and g are regular within and on C. Consequently,

$$(z^n, z^m) = \frac{1}{2i(m+1)} \int_{|z|=1} z^n \bar{z}^{m+1}\, dz.$$
$$= 0.$$

A similar calculation with $m = n$ yields the normalising factor.

(ii) Completeness We shall verify Parseval's theorem
directly. Green's theorem is used again, this time to find an
expression for the Fourier coefficients (c_n) for $f \in L_r^2(D)$:

$$c_n = \lim_{r \to 1} \sqrt{\left(\frac{n}{\pi}\right)} \iint_{|z|<r} f(z)\,\bar{z}^{n-1}\, dx\, dy$$
$$= \lim_{r \to 1} \frac{1}{2i} \sqrt{\left(\frac{n}{\pi}\right)} \int_{|z|=r} f(z)\,\frac{\bar{z}^n}{n}\, dx$$
$$= \lim_{r \to 1} \frac{1}{\sqrt{(\pi n)}} \frac{r^{2n}}{2i} \int_{|z|=r} \frac{f(z)}{z^n}\, dz.$$

Now the coefficients (b_n) in the power series expansion for f are
given by the formula

$$b_{n-1} = \frac{1}{2\pi i} \int_{|z|=r} \frac{f(z)}{z^n}\, dz, \quad r < 1,$$

hence $\quad c_n = \sqrt{\left(\frac{\pi}{n}\right)}\, b_{n-1} \quad (n = 1, 2, \ldots).$ $\hspace{1cm}$ (5)

Next, by formal calculations, we have

$$\|f\|^2 = \iint_D |f(z)|^2 \, dx \, dy$$

$$= \int_0^1 \int_0^{2\pi} \left\{ \sum_{n=0}^{\infty} b_n \, r^n \, e^{in\theta} \right\} \left\{ \sum_{n=0}^{\infty} \bar{b}_n \, r^n \, e^{-in\theta} \right\} r \, dr \, d\theta$$

$$= \int_0^1 \int_0^{2\pi} \sum_{n=0}^{\infty} \sum_{m=0}^{\infty} b_n \, \bar{b}_m \, r^{n+m} \, e^{i(n-m)\theta} \, r \, dr \, d\theta$$

$$= \int_0^1 \sum_{n=1}^{\infty} \sum_{m=0}^{\infty} b_n \, \bar{b}_m \int_0^{2\pi} e^{i(n-m)\theta} \, d\theta \, r^{n+m+1} \, dr$$

$$= 2\pi \int_0^1 \sum_{n=0}^{\infty} |b_n|^2 \, r^{2n+1} \, dr$$

$$= \pi \sum_{n=0}^{\infty} \frac{|b_n|^2}{n+1}.$$

These formalities can be justified by considerations of uniform and of absolute convergence; then because of (5) we have

$$\|f\|^2 = \pi \sum_{n=1}^{\infty} \frac{|b_{n-1}|^2}{n} = \sum_{n=1}^{\infty} |c_n|^2$$

and the Parseval relation is verified.

For any region R of the complex plane the Hilbert space $L_r^2(R)$ is an r.k. space, since all its members are continuous (corollary, p. 54). It was in the context of conformal mapping of the underlying region R that the reproducing kernel was first introduced by Stefan Bergman. Indeed, kernels of this type are often called Bergman kernels; they have important connections with conformal mapping functions of the region R and with Green's kernel for R.

Since we now have a CON sequence for $L_r^2(D)$, we can calculate the reproducing kernel $k_D(z, \zeta)$, using the formula of property (e) of § 2.5:

$$k_D(z, \zeta) = \sum_{n=0}^{\infty} \frac{(n+1)}{\pi} z^n \bar{\zeta}^n$$

$$= \frac{1}{\pi} \frac{d}{d(z\bar{\zeta})} \sum_{n=0}^{\infty} (z\bar{\zeta})^{n+1}$$

$$= \frac{1}{\pi} \frac{d}{d(z\bar{\zeta})} \frac{z\bar{\zeta}}{1 - z\bar{\zeta}},$$

so that $k_D(z, \zeta) = 1/(1 - z\bar{\zeta})^2$.

We now investigate the behaviour of CON sequences for $L_r^2(R)$ under conformal mapping of the region R.

Let $w(\zeta)$ be regular in a simply connected domain R of the $\zeta = \xi + i\eta$ plane, $w'(\zeta) \neq 0$ in R, and let w map R conformally onto R', a region in the z-plane. Then $\zeta = p(z) = w^{-1}$ is a conformal map of R' onto R and

$$\frac{dw}{d\zeta}\frac{dp}{dz} = 1.$$

If $g \in L_r^2(R)$,

$$\iint_{R'} |g(z)|^2 \, dx \, dy = \iint_R |g(w(\zeta))|^2 |w'(\zeta)|^2 \, d\xi \, d\eta,$$

$|w'|^2$ being the Jacobian of the mapping. In terms of norms, this is

$$\|g\|_{R'} = \|(g \circ w) \, w'\|_R.$$

DEFINITION   The association

$$g \leftrightarrow (g \circ w) \, w'$$

is a mapping between $L_r^2(R)$ and $L_r^2(R')$, which we shall call *the mapping induced by w*, or just *the induced mapping*.

As we have just seen the induced mapping is norm preserving, and we wish to apply the ideas of § 2.6.1 to it. In fact, *the induced mapping is an isometric isomorphism between $L_r^2(R)$ and $L_r^2(R')$.* To complete the proof of this statement, we note that since the induced mapping is linear and norm preserving, and hence one-to-one from $L_r^2(R')$ into $L_r^2(R)$, we have only to show that it is also 'onto'. Consequently, let $F \in L_r^2(R)$ and set $g(z) = F(p(z)) \, p'(z)$ (recall that $p = w^{-1}$). Then

$$\iint_R |F(\zeta)|^2 \, d\xi \, d\eta = \iint_R |F(p(w(\zeta))) \, p'(w(\zeta)) \, w'(\zeta)|^2 \, d\xi \, d\eta$$

$$= \iint_R |g(w(\zeta)) \, w'(\zeta)|^2 \, d\xi \, d\eta$$

$$= \iint_{R'} |g(z)|^2 \, dx \, dy < \infty.$$

That is, $g \in L_r^2(R')$ as required.

As an example of the foregoing remarks, we can generate a CON sequence for $L_r^2(R)$ where R is any simply connected domain by mapping it onto the unit disc. If w is the mapping function, the induced mapping associates

$$\left\{ \sqrt{\left(\frac{n}{\pi}\right)} \, \zeta^{n-1} : n = 1, 2, \ldots \right\},$$

forming a CON sequence in $L_r^2(D)$ with

$$\left\{ \sqrt{\left(\frac{n}{\pi}\right)} \, [w(z)]^{n-1} \, w'(z) : n = 1, 2, \ldots \right\},$$

forming a CON sequence in $L^2(R)$.

In particular, we can generate another CON sequence for $L_r^2(D)$ by mapping D onto itself by the linear fractional transformation

$$w = \frac{z-a}{1-\bar{a}z}, \quad |a| < 1.$$

Now

$$\frac{dw}{dz} = \frac{1-|a|^2}{(1-\bar{a}z)^2}$$

so that another CON sequence is formed by the set

$$\left\{ \sqrt{\left(\frac{n}{\pi}\right)} \left(\frac{z-a}{1-\bar{a}z}\right)^{n-1} \frac{1-|a|^2}{(1-\bar{a}z)^2} : n = 1, 2, \ldots \right\}.$$

Next we state a completeness theorem for complex polynomials.

THEOREM   *If R is a finite region of the complex plane, whose complement is a closed region, then the set of powers $\{z^n : n = 0, 1, \ldots\}$ forms a complete sequence in $L_r^2(R)$.*

The proof is beyond our present scope; it may be found in Nehari (1952). As with the completeness theorem for real polynomials (p. 31), this theorem applies to any simple set of polynomials defined on R; in particular, to the ON sequence formed by orthonormalising (z^n) over R.

We follow this with a remarkable

EXAMPLE   The Chebyshev polynomials of the second kind, $\{U_n(z)\}$ (Appendix 2), form a CON sequence for $L_r^2(E)$, where E is the interior of the ellipse $b^2 x^2 + a^2 y^2 = ab$.

Proof The previous theorem gives the completeness, since $\{U_n\}$ is a simple set. As for the orthogonality, we can avoid the necessity of orthonormalising (z^n) over E by using a conformal mapping technique. The function $w = \cos^{-1}z$ maps the region E which has been cut from its foci to the boundary along the real axis conformally on the rectangular region $R = \{w: 0 < u < \pi, |v| < c\}$ in the $w = u + iv$ plane, where $a = \cosh c$ and $b = \sinh c$.

We use the representation (Magnus *et al.* (1966) p. 257)

$$U_n(z) = (1 - z^2)^{-\frac{1}{2}} \sin\left[(n+1)\cos^{-1}z\right]$$

and the fact that the cuts do not affect the value of the inner product

$$(U_n, U_m) = \iint_E U_n(z)\, \overline{U_m(z)}\, dx\, dy.$$

Thus from the mapping of $L_r^2(E) \to L_r^2(R)$ induced by w, we have

$$(U_n, U_m) = \iint_R \sin(n+1)\,w \sin(m+1)\,w\, du\, dv$$

which, by the addition theorem for sine and by symmetry considerations, is readily found to be zero. Similarly, the normalising factor is found, and the result is that

$$\left\{2\left[\frac{n+1}{(A^{n+1} - A^{-n-1})}\right]^{\frac{1}{2}} U_n(z) : n = 0, 1. ...\right\},$$

where $A = (a+b)^2$, forms a CON sequence in $L^2(E)$.

Problems

2.29 By using suitable mapping functions, find CON sequences for $L_r^2(R)$, where R is (a) the interior of the circle $|z - a| = r$; (b) the upper half-plane; (c) the interior of a square. Find CON sequences for $L_r^2(D)$ and $L_r^2(E)$ which have not already been derived in the text.

2.30 Let $L^2(D)$ denote the full L^2 space over the unit disc. Find a CON sequence for $L^2(D)$ and use the circular form of Vitali's criterion (p. 38) to prove the completeness. Find a CON sequence for $\bar{L}_r^2(D)$, the class of L^2 functions whose complex conjugates are regular in D, and show that an element of $\bar{L}_r^2(D)$ can only have a constant component in common with an element of $L_r^2(D)$. Find an element of $L^2(D)$ which is orthogonal to $\bar{L}_r^2(D) \cup L_r^2(D)$.

2.31 Show that
$$\left\{ \frac{z^n}{\sqrt{(\pi n(1 - r^{2n}))}} : n = \pm 1, \pm 2, \dots \right\},$$

forms a CON sequence for $L_r^2(A)$, where A is the annular region $\{z : 0 < r < |z| < 1\}$. The reader must take careful note that this is the first non-simply connected region that has been introduced.

3. Non-orthogonal sequences

In this chapter we shall give some methods for demonstrating completeness and basis properties of sequences which do not have the property of orthogonality. Thus we are forced to give up the very useful Parseval relation, and other types of criteria must be sought. The most useful of these involves the idea of 'stability', and we shall give a brief introduction to this topic in § 3.1.

3.1 The stability of bases

We shall find that basis properties of sequences can sometimes be deduced from the fact that they are 'near' in some sense to a sequence already known to possess the required property. We list some types of nearness:

   (i) (ψ_n) is *strictly near* (ϕ_n) if $\Sigma \|\phi_n - \psi_n\| = \lambda < \infty$.

   (ii) (ψ_n) is *quadratically near* (ϕ_n) if $\Sigma \|\phi_n - \psi_n\|^2 = \lambda < \infty$.

   (iii) (ψ_n) is *KL (Krein–Lyusternik) near* the biorthogonal system $\{\phi_n, \phi_n^*\}$ if
$$\Sigma \|\phi_n - \psi_n\| \, \|\phi_n^*\| = \lambda < \infty.$$

   (iv) (ψ_n) is *Paley–Wiener near* (ϕ_n) if
$$\|\Sigma a_n(\phi_n - \psi_n)\| < \lambda \|\Sigma a_n \phi_n\|,$$
for all finite sequences of scalars (a_n), and for some $\lambda < 1$.

DEFINITION A property of a sequence (ϕ_n) in Banach space is said to be *stable* if it conserved by all sequences sufficiently near to it in some sense.

Our main goal in this section is the Paley–Wiener stability theorem; from it we will be able to deduce, in § 3.1.1, some properties of the non-harmonic Fourier functions. Before this, however, we shall give some other typical stability theorems (A, B, C and D below); theorem A will be found useful later in the context of differential operators (see § 4.1).

THEOREM A   *Let (ϕ_n) be a* CON *sequence in a Hilbert space H, and let (ψ_n) be quadratically near to (ϕ_n). Then (ψ_n) is complete in H.*

Proof   Choose N so large that

$$\sum_{n=N+1}^{\infty} \|\phi_n - \psi_n\|^2 < 1.$$

Now let us subtract from ϕ_n its component along the ψ_ns for $n > N$, i.e. set

$$\eta_n = \phi_n - \sum_{k=N+1}^{\infty} (\phi_k, \psi_k)\, \psi_k.$$

In the first part of the proof we show that the set

$$\{\eta_n : n = 1, ..., N\} \cup \{\psi_n : n = N+1, ..., \infty\}$$

forms a complete sequence in H. Suppose then that for $h \in H$, $h \neq \theta$, we have

$$(h, \eta_n) = 0, \quad n = 1, 2, ..., N$$

and

$$(h, \psi_n) = 0, \quad n = N+1, ...$$

Then h is also orthogonal to ϕ_n $(n = 1, 2, ..., N)$ since

$$0 = (h, \eta_n) = (h, \phi_n) - \sum_{k=N+1}^{\infty} (\phi_n, \psi_k)(h, \psi_k) = (h, \phi_n).$$

Using the Parseval relation, we have

$$\|h\|^2 = \sum_{k=1}^{\infty} |(h, \phi_k)|^2 = \sum_{k=N+1}^{\infty} |(h, \phi_k)|^2$$

$$= \sum_{k=N+1}^{\infty} |(h, \phi_k - \psi_k)|^2$$

$$\leqslant \sum_{k=N+1}^{\infty} \|h\|^2 \|\phi_k - \psi_k\|^2$$

$$< \|h\|^2 \text{ by the choice of } N.$$

This contradiction shows that the assumption $h \neq \theta$ is false, and completes the first part.

Secondly, set $S = [\psi_{N+1}, \ldots]^{\perp}$. Then $\eta_n \in S$ $(n = 1, \ldots, N)$ by construction; in fact by the first part S is the closed linear span of $\{\eta_n : n = 1, \ldots, N\}$ and so is of dimension N. But the mutually orthogonal elements ψ_1, ψ_2, ..., ψ_N also lie in S, and therefore span S. Then $\eta_1, \ldots, \eta_N$ must be linear combinations of $\psi_1, \ldots, \psi_N$, and

$$[\psi_1, \ldots, \psi_N, \psi_{N+1}, \ldots]^{\perp} = [\eta_1, \ldots, \eta_N, \psi_{N+1}, \ldots]^{\perp}$$
$$= [\theta]$$

by the first part. In other words, (ψ_n) is complete in H.

Next we give without proof some stability theorems which are similar to the last one, but more powerful. For these we need the

DEFINITION   A sequence (ϕ_n) in Banach space is called ω-linearly independent if

$$\sum_{k=1}^{\infty} \phi_k \alpha_k = \theta \Rightarrow \alpha_k = 0 \text{ for every } k.$$

This evidently generalises the notion of linear independence in finite dimensional vector space.

THEOREM B   (Kato (1966) p. 265) *Let (ϕ_n) be a CON sequence in a Hilbert space H, and (ψ_n) an ω-linearly independent sequence in H that is quadratically near (ϕ_n). Then (ψ_n) is a basis for H.*

If $\sum_n \|\phi_n - \psi_n\|^2 < 1$, then (ψ_n) is already ω-linearly independent.

THEOREM C   (Kato (1966) p. 266) *If (ψ_n) is strictly near a CON sequence in Hilbert space, with $\lambda < 1$, then it is a basis.*

THEOREM D   (Singer (1970) p. 94) *Let (ψ_n) be a sequence in a Banach space B which is KL near a basis $\{\phi_n, \phi_n^*\}$ for B. Then (i) (ψ_n) is a basis if and only if it is ω-linearly independent; (ii) (ψ_n) is a basis if and only if it is total in B.*

In preparation for our study of the Paley–Wiener stability theorem we introduce certain special types of basis for Banach

space, described in the following definition; note that a CON
sequence in Hilbert space is a basis of each of the three types
listed, and the situation can be viewed as another illustration of
the favourable properties enjoyed by CON sequences which may
be lost along with orthogonality.

DEFINITION   Let (ϕ_n) be a basis for a Banach space B. Then
(i) (ϕ_n) is called a *Bessel basis* if

$$\Sigma\, a_k\, \phi_k \ \text{convergent} \Rightarrow \Sigma |a_k|^2 < \infty,$$

or equivalently, if for some constant $C > 0$,

$$\sum_{}^{N} |a_k|^2 \leqslant C \,\Big\|\sum_{}^{N} a_k\, \phi_k\Big\|^2$$

for all finite sequences (a_n) of scalars.
(ii) (ϕ_n) is called a *Hilbert basis* if

$$\Sigma |a_k|^2 < \infty \Rightarrow \Sigma\, a_k\, \phi_k \ \text{convergent},$$

or equivalently, if for some constant $C' > 0$,

$$\Big\|\sum_{}^{N} a_k\, \phi_k\Big\|^2 \leqslant C' \sum_{}^{N} |a_k|^2$$

for all finite sequences (a_n) of scalars.
(iii) (ϕ_n) is called a *Riesz basis* or *Riesz–Fischer basis* if it is
both a Bessel and a Hilbert basis, or equivalently if there exist
constants A and B, $0 < A \leqslant B < \infty$ such that

$$A\{\sum_{}^{N} |a_k|^2\}^{\frac{1}{2}} \leqslant \Big\|\sum_{}^{N} a_k\, \phi_k\Big\| \leqslant B\{\sum_{}^{N} |a_k|^2\}^{\frac{1}{2}}$$

for all finite sequences (a_n) of scalars.
Proofs of the equivalences can be found in Singer (1970) p. 338.

DEFINITION   We call two sequences (ϕ_n) and (ψ_n) in Banach
space B *equivalent* if there exists a bounded invertible operator
T on B such that $\psi_n = T(\phi_n)$ for every n.

Note that by 'invertible' we mean that T^{-1} exists and is
bounded on all of B. As well as nearness in the various senses
just described, equivalence is another way of linking two sets in
B. The main reason for introducing equivalence is to be found
in the following

LEMMA   *If a sequence (ψ_n) in a Banach space is equivalent to a basis (ϕ_n), it too is a basis.*

Proof   Let $T : B \to B$, $T(\phi_n) = \psi_n$ for every n and T^{-1} exist and be bounded on all of B. Let $f \in B$.

(i) $\left\| f - \overset{N}{\Sigma} a_n \psi_n \right\| = \left\| f - \overset{N}{\Sigma} a_n (T\phi_n) \right\|$

$$= \left\| TF - \overset{N}{\Sigma} a_n (T\phi_n) \right\| \text{ for some } F \in B,$$

$$= \left\| T(F - \overset{N}{\Sigma} a_n \phi_n) \right\|$$

$$\leqslant \| T \| \left\| F - \overset{N}{\Sigma} a_n \phi_n \right\|.$$

This last expression tends to 0 as N tends to ∞ if the a_n are chosen to be the Fourier coefficients of F with respect to (ϕ_n). Thus each $f \in B$ has an expansion in the set $\{\psi_n\}$.

(ii) The coefficients are unique, since if (a'_n) is any sequence of coefficients such that $\| f - \Sigma a'_n \psi_n \| \to 0$ and (a_n) is as in (i), then

$$\left\| \Sigma (a'_n - a_n) \psi_n \right\| \leqslant \| f - \Sigma a'_n \psi_n \| - \| f - \Sigma a_n \psi_n \| \to 0.$$

Thus

$$\left\| \overset{N}{\Sigma} (a'_n - a_n) \phi_n \right\| = \left\| T^{-1} \overset{N}{\Sigma} (a'_n - a_n) \psi_n \right\|$$

$$\leqslant \| T^{-1} \| \left\| \overset{N}{\Sigma} (a'_n - a_n) \psi_n \right\| \to 0 \text{ as } N \to \infty,$$

and since (ϕ_n) is a basis we must have $a'_n = a_n$ for every n, and the proof is complete.

Gelfand showed, for example, that a basis of Hilbert space is absolute if and only if it is equivalent to a CON sequence.

Problems

3.1   Give the finite dimensional vector space analogues of the previous lemma and of the result (p. 55) asserting the preservation of the CON property under unitary transformation of Hilbert space.

3.2   Show that the properties of being a Bessel, Hilbert or Riesz basis are preserved under equivalence.

THEOREM (Paley–Wiener stability theorem)   *Let* (ϕ_n) *be a basis for a Banach space B and* (ψ_n) *a sequence in B. Suppose that*

$$\left\| \sum_n a_n(\phi_n - \psi_n) \right\| \leqslant \lambda \left\| \sum_n a_n \phi_n \right\| \tag{1}$$

where $0 < \lambda < 1$ *and* (a_n) *is any finite sequence of scalars; that is,* (ψ_n) *is Paley–Wiener near* (ϕ_n). *Then* (ψ_n) *is also a basis for B.*

Note that, because of

$$\left\| \Sigma a_n(\phi_n - \psi_n) \right\| \geqslant \left| \left\| \Sigma a_n \phi_n \right\| - \left\| \Sigma a_n \psi_n \right\| \right|$$

the hypothesis (1) implies

$$(1-\lambda)\|\Sigma a_n \phi_n\| \leqslant \|\Sigma a_n \psi_n\| \leqslant (1+\lambda)\|\Sigma a_n \phi_n\|. \tag{2}$$

We note also that since

$$\|\Sigma a_n \phi_n\| = \{\Sigma |a_n|^2\}^{\frac{1}{2}}$$

if B is a Hilbert space and (ϕ_n) a CON sequence, we can, using the definition of a Riesz basis, state the following useful form of the Paley–Wiener stability theorem:

COROLLARY   *Let H be a Hilbert space with CON sequence* (ϕ_n) *and let* (ψ_n) *be a sequence in H. If*

$$\|\Sigma a_n(\phi_n - \psi_n)\| \leqslant \lambda\{\Sigma |a_n|^2\}^{\frac{1}{2}}$$

with $\lambda < 1$, *for all finite sequences of scalars* (a_n), *then* (ψ_n) *is a Riesz basis for H.* There are many other similar theorems; see e.g. Singer (1970).

Proof of the Paley–Wiener stability theorem.   The method of proof is to show that, under the hypotheses, (ψ_n) is equivalent to (ϕ_n) in the sense of the previous lemma. To do this we introduce the operator
$$T : \Sigma a_k \phi_k \to \Sigma a_n \psi_n,$$

where (a_n) is any finite sequence of scalars. We can extend T to all of B 'by continuity' (Appendix 1,8), since if $(F_M) = (\overset{M}{\Sigma} a_n \phi_n)$ converges in B as $M \to \infty$, then $(G_M) = (T(F_M))$ converges, since by (2) $(\overset{M}{\Sigma} a_n \psi_n)$ is a Cauchy sequence.

Also from (2) we find that

$$\|T(\overset{M}{\underset{}{\Sigma}} a_n \phi_n)\| = \|\overset{M}{\underset{}{\Sigma}} a_n \psi_n\| \leqslant (1+\lambda)\|\overset{M}{\underset{}{\Sigma}} a_n \phi_n\|$$

from which it follows that T as extended is bounded.

Obviously we have $T(\phi_n) = \psi_n$ for every n, so to establish equivalence it remains to be shown that T is invertible. If I denotes the identity operator on B we have

$$\|(I-T)\Sigma a_n \phi_n\| = \|\Sigma a_n(\phi_n - \psi_n)\| \leqslant \lambda\|\Sigma a_n \phi_n\|$$

therefore $\|I-T\| \leqslant \lambda < 1$, i.e. T is 'sufficiently close to I' to be invertible (see Appendix 1,9). This completes the proof.

We may remark that the key to this proof was the fact that the closeness of (ψ_n) to (ϕ_n) in the Paley–Wiener sense led to the fact that T was sufficiently close to the identity to be invertible, leading in turn to the equivalence of (ψ_n) and (ϕ_n). This proof, using as it does the theory of operators, makes an interesting contrast to the proof of the 'quadratic' stability theorem (A) which used only the geometry of Hilbert space in its arguments; so also did Paley's and Wiener's original proof of their stability theorem (which had the more modest aim of showing among other things, that if the ON sequence (ϕ_n) is complete in L^2 then so is (ψ_n)); this proof is to be found in Paley and Wiener (1934) p. 100.

In order to make practical use of the theorem we must be able to verify the hypothesis (1) for particular sequences (ϕ_n) and (ψ_n). Duffin and Eachus (1942) have given a method of 'separation of variables' suitable for application to Hilbert space, and the lemma which follows is a slightly modified version of their method; it is well suited to the study of the non-harmonic Fourier functions, as we shall see in the next section.

LEMMA   *Let (ϕ_n) be a CON sequence in a Hilbert space H, and (ψ_n) a sequence in H. Suppose that*

$$\phi_n - \psi_n = \overset{\infty}{\underset{k=1}{\Sigma}} C_{n,k}(T_k \phi_n) \tag{3}$$

strongly, where $(C_{n,k})$ is a matrix of scalars with $|C_{n,k}| < c_k$, and (T_k) is a sequence of bounded linear operators on H with $\|T_k\| < t_k$.

Then (ψ_n) *is Paley–Wiener near* (ϕ_n) *if* $\sum\limits_{k=1}^{\infty} c_k t_k < 1$, *and consequently* (ψ_n) *is a Riesz basis for* H.

Proof Since the norm is continuous on H to $\mathbb{R}$, we have from (3)

$$\left\|\sum_n a_n(\phi_n - \psi_n)\right\| = \lim_{N\to\infty} \left\|\sum_n a_n \sum_{k=1}^{N} C_{n,k} T_k \phi_n\right\|.$$

Now

$$\left\|\sum_n a_n \sum_{k=1}^{N} C_{n,k} T_k \phi_n\right\|$$

$$= \left\|\sum_{k=1}^{N} T_k \sum_n a_n C_{n,k} \phi_n\right\|$$

$$\leqslant \sum_{k=1}^{N} \left\|T_k \sum_n a_n C_{n\,k} \phi_n\right\|$$

$$\leqslant \sum_{k=1}^{N} t_k \left\|\sum_n C_{n,k} a_n \phi_n\right\|$$

$$\leqslant \sum_{k=1}^{N} t_k c_k \{\textstyle\sum_n |a_n|^2\}^{\frac{1}{2}},$$

the last inequality being true because, if the orthogonality of (ϕ_n) be used twice, we get

$$\left\|\sum_n C_{n,k} a_n \phi_n\right\|^2 = \sum_n \|C_{n,k} a_n \phi_n\|^2$$

$$\leqslant |c_k|^2 \sum_n |a_n|^2.$$

We now have

$$\left\|\sum_n a_n(\phi_n - \psi_n)\right\| \leqslant \left(\lim_{N\to\infty} \sum_{k=1}^{N} t_k c_k\right) \{\textstyle\sum_n |a_n|^2\}^{\frac{1}{2}}$$

and the hypotheses of the corollary to the Paley–Wiener stability theorem are satisfied with $\lambda = \sum\limits_k t_k c_k$. This completes the proof.

3.1.1   The non-harmonic Fourier functions The functions $\{e^{i\lambda_n x}: n = 0, \pm 1, \pm 2, \ldots\}$ are called the *non-harmonic Fourier functions*, or sometimes just *complex exponentials*; (λ_n) is a given sequence of complex numbers. The study of these functions was initiated by N. Wiener, and it was to them that Paley and Wiener applied their original stability theorem. They obtained the result that if the λ_n are real and such that

$$|\lambda_n - n| \leqslant D < \pi^{-2}$$

then $\{e^{i\lambda_n x}: n = 0, \pm 1, \ldots\}$ forms a complete sequence in $L^2(-\pi, \pi)$. Later Levinson (1936a) improved the result by enlarging the constant π^{-2} to $\frac{1}{4}$. He treated the case of $L^p(-\pi, \pi)$, $1 < p \leqslant 2$, and showed that the best possible value for D is $(p-1)/2p$, in the sense that completeness holds on $L^p(-\pi, \pi)$ if

$$|\lambda_n - n| \leqslant D < (p-1)/2p$$

but may fail to hold if $|\lambda_n - n| \leqslant (p-1)/2p$.

The case of complex λ_n was investigated by Duffin and Eachus (1942) using the separation of variables method. This is the content of our next

EXAMPLE   Let (λ_n) be a sequence of complex numbers such that $|\lambda_n - n| \leqslant D < (\log 2)/\pi$. Then

$$\{e^{i\lambda_n x}: n = 0, \pm 1, \ldots\} \text{ forms a Riesz basis for } L^2(-\pi, \pi).$$

We have
$$e^{inx} - e^{i\lambda_n x} = (1 - e^{i(\lambda_n - n)x})e^{inx}$$

$$= -\sum_{k=1}^{\infty} \frac{[i(\lambda_n - n)x]^k}{k!} e^{inx}$$

strongly in $L^2(-\pi, \pi)$. With the notations of the previous lemma we find that T_k is the operation of multiplication by x^k, and the norm of this operator is easily calculated to be π^k. Furthermore, $C_{n,k}! = -[i(\lambda_n - n)]^k/k!$, and if we assume that $|\lambda_n - n| \leqslant D$ then $c_k = D^k/k!$. Thus

$$\sum_{k=1}^{\infty} c_k t_k = e^{\pi D} - 1.$$

If this is to be less than 1 we shall have $e^{\pi D} < 2$, or $D < (\log 2)/\pi$.

Problem

3.3 Show that $\{e^{i\lambda_n x}: n = 0, \pm 1, \ldots\}$ is orthogonal over $(-\pi, \pi)$ if and only if $\lambda_n = n + a$, a real. Show that this orthogonal set forms a complete sequence in $L^2(-\pi, \pi)$. Now let λ_n be complex; show that $\{e^{i\lambda_n x}\}$ forms a Riesz basis for $L^2(-\pi, \pi)$

if $\lambda_n - n$ lies in the closure of any region interior to a circle in the $x + iy$ plane with real centre and radius $(\log 2)/\pi$. In particular, if λ_n is real, $\lambda_n - n$ is of the same sign for every n, and $|\lambda_n - n| \leqslant D < (2\log 2)/\pi$ then $\{e^{i\lambda_n x}\}$ forms a Riesz basis for $L^2(-\pi, \pi)$.

If $|\lambda_n - n| \leqslant D$ and each λ_n is real we have seen that $D < \frac{1}{4}$ is sufficient for $\{e^{i\lambda_n x}\}$ to form a complete sequence in $L^2(-\pi, \pi)$, and that $D < (\log 2)/\pi$ is sufficient for $\{e^{i\lambda_n x}\}$ to form a Riesz basis for $L^2(-\pi, \pi)$, even for complex λ_n. It is natural to ask whether $(\log 2)/\pi$ (which equals 0.22 approximately) can be improved. It turns out that for real λ_n this constant can indeed be enlarged to $\frac{1}{4}$, the best possible value, with no consequent loss of the Riesz basis property. This remarkable fact was discovered by M. I. Kadec (1964) again using the separation of variables method as follows:

EXAMPLE   For real λ_n, $\{e^{i\lambda_n x}: n = 0, \pm 1, \ldots\}$ forms a Riesz basis for $L^2(-\pi, \pi)$ if $|\lambda_n - n| \leqslant D < \frac{1}{4}$.

We proceed as in the previous example but, instead of expanding $1 - e^{i(\lambda_n - n)x}$ into power series, we expand in the set

$$\left\{ \frac{1}{\sqrt{(2\pi)}}, \frac{\cos kx}{\sqrt{\pi}}, \frac{\sin(k - \frac{1}{2})x}{\sqrt{\pi}} \right\}$$

(see p. 37) over $(-\pi, \pi)$. With $\lambda_n - n = \mu_n$ we obtain

$$1 - e^{i\mu_n x} = 1 - \frac{\sin \pi \mu_n}{\pi \mu_n} + \sum_{k=1}^{\infty} \frac{(-1)^k 2\mu_n \sin \pi \mu_n}{\pi(k^2 - \mu_n^2)} \cos kx$$

$$+ i \sum_{k=1}^{\infty} \frac{(-1)^k 2\mu_n \cos \pi \mu_n}{\pi[(k - \frac{1}{2})^2 - \mu_n^2]} \sin(k - \frac{1}{2})x.$$

Now let T_0, T_{2k+1}, T_{2k}, represent multiplication by 1, $\cos kx$ and $\sin(k - \frac{1}{2})x$ respectively.

Also let
$$C_{n,0} = 1 - \frac{\sin \pi \mu_n}{\pi \mu_n},$$

$$C_{n,2k+1} = \frac{(-1)^k 2\mu_n \sin \pi \mu_n}{\pi(k^2 - \mu_n^2)},$$

$$C_{n,2k} = \frac{i(-1)^k 2\mu_n \cos \pi \mu_n}{\pi[(k - \frac{1}{2})^2 - \mu_n^2]}.$$

With the notations as in the previous lemma we have

$$t_0 = t_{2k+1} = t_{2k} = 1,$$

$$c_0 = 1 - \frac{\sin \pi D}{\pi D},$$

$$c_{2k+1} = \frac{2D \sin \pi D}{\pi(k^2 - D^2)},$$

$$c_{2k} = \frac{2D \cos \pi D}{\pi[(k - \frac{1}{2})^2 - D^2]},$$

provided that D is suitably restricted, certainly for $0 < D < \frac{1}{4}$. Thus

$$\sum_{k=0}^{\infty} c_k t_k = 1 - \frac{\sin \pi D}{\pi D} + \sin \pi D \sum_{k=1}^{\infty} \frac{2D}{\pi(k^2 - D^2)}$$

$$+ \cos \pi D \sum_{k=1}^{\infty} \frac{2D}{\pi[(k - \frac{1}{2})^2 - D^2]}.$$

The two series on the right-hand side are the well-known expansions for $(\pi D)^{-1} - \cot \pi D$ and for $\tan \pi D$ respectively, so that

$$\sum_{k=0}^{\infty} c_k t_k = 1 - \cos \pi D + \sin \pi D$$

$$< 1$$

if $D < \frac{1}{4}$, as required.

An extension to the case of complex λ_n was also given by Kadec, but Young (1975) has pointed out an error in the proof; Young gives the following result:

Let $\lambda_n = \lambda_n' + i\lambda_n''$, and C and D be constants such that

$$|\lambda_n' - n| \leqslant D < \tfrac{1}{4},$$

$$|\lambda_n''| \leqslant C < \infty.$$

Then $\{e^{i\lambda_n x} : n = 0, \pm 1, \ldots\}$ forms a basis for $L^2(-\pi, \pi)$. Young's proof shows that this basis is in fact Riesz.

There follow some examples of the stability theorems on pp. 72–3 applied to the non-harmonic Fourier functions with real λ_n. To apply these theorems we shall need

$$(2\pi)^{-\frac{1}{2}} \| e^{inx} - e^{i\lambda_n x} \|_{L^2} = (2\pi)^{-\frac{1}{2}} \left\{ \int_{-\pi}^{\pi} |e^{inx} - e^{i\lambda_n x}|^2 \, dx \right\}^{\frac{1}{2}}$$

$$= 2^{\frac{1}{2}} \left\{ \frac{1 - \sin \pi(\lambda_n - n)}{\pi(\lambda_n - n)} \right\}^{\frac{1}{2}}$$

$$= (2S_n)^{\frac{1}{2}}, \text{ say.}$$

We can now make the following deductions:

(1) From theorem A, if $\sum_{-\infty}^{\infty} S_n < \infty$, then $(e^{i\lambda_n x})$ is complete in $L^2(-\pi, \pi)$.

(2) $(e^{i\lambda_n x})$ is a basis for $L^2(-\pi, \pi)$ if either $\sum_{-\infty}^{\infty} S_n < \frac{1}{2}$, from theorem B, or $\sum_{-\infty}^{\infty} S_n^{\frac{1}{2}} < 2^{-\frac{1}{2}}$, from theorem C.

(3) From theorem D, *if* $\sum_{-\infty}^{\infty} S_n^{\frac{1}{2}} < \infty$, *then* $(e^{i\lambda_n x})$ *is a basis for* $L^2(-\pi, \pi)$ *if it is complete.*

Sequences (λ_n) which validate the criteria in (1) or (3) above can easily be found, by taking account of $1 + X^{-1} \sin X = O(X^2)$ as $X \to 0$. Thus, with $\lambda_n - n = \mu_n$ we may take $\mu_n = O(n^{-\alpha})$, $n \to \infty$, $\alpha > \frac{1}{2}$, as a sufficient condition for the convergence in (1), and $\mu_n = O(n^{-\alpha})$, $n \to \infty$, $\alpha > 1$, for the convergence in (3). Further to the result for (3), we note that if $\mu_n = O(n^{-\alpha})$, $n \to \infty$, $\alpha > 1$, then $(e^{i\lambda_n x})$ is complete in $L^2(-\pi, \pi)$. This can be deduced from two results of Levinson, one of which asserts that $(e^{i\lambda_n x})$ is complete in $L^2(-\pi, \pi)$ if $|\mu_n| \leqslant D < \frac{1}{4}$, as we have already observed, the other (Levinson (1936 b)) asserting that a sequence $(e^{i\lambda_n x})$ which is total in $L^p(-\pi, \pi)$ remains total if any one λ_n is replaced with some other number. (By using problem 1.8 (i), we can replace 'total' in this last assertion with 'complete'.)

To see this we observe that given any D such that $0 < D < \frac{1}{4}$, there exists N such that $\{\lambda_n\}$ satisfies $|\mu_n| \leqslant D$ for all n such that $|n| > N$. Consider the set

$$\{\lambda_n : |n| > N\} \cup \{n : |n| \leqslant N\}.$$

This set satisfies $|\mu_n| \leqslant D$ for every n, therefore

$$\{e^{i\lambda_n x} : |n| > N\} \cup \{e^{inx} : |n| \leqslant N\}$$

forms a complete sequence in $L^2(-\pi, \pi)$. We now replace each

n in the second member with λ_n, obtaining the complete sequence $(e^{i\lambda_n x})_{-\infty < n < \infty}$.

These remarks can now be combined with the result (3) above to obtain: $(e^{i\lambda_n x})$ *is a basis for* $L^2(-\pi,\pi)$ *if* $\lambda_n - n = O(n^{-\alpha})$, $n \to \infty$, $\alpha > 1$.

Problems

3.4 Use similar arguments, involving the more general Levinson L^p-completeness criterion (p. 79) and Hölder's inequality to show that if $\mu_n = O(n^{-\alpha})$, $n \to \infty$, $\alpha > 1/p$, then $(e^{i\lambda_n x})$ is a basis for $L^p(-\pi,\pi)$.

3.5 Show that the first criterion of deduction (2) above is weaker than the second, but stronger than the criterion $|\lambda_n - n| \leqslant D < \frac{1}{4}$ of the previous example.

It is interesting to note the difference between the type of criterion of this last result and that of the Kadec result (p. 80) deduced from the Paley–Wiener stability theory. Instead of requiring all the λ_n to be reasonably close to n, in the present discussion the first finitely many λ_n can be any numbers whatever, but this must be compensated for by requiring that $\lambda_n \to n$ in a prescribed manner as $n \to \infty$.

As a postscript to this section, let us observe that the quantity S_n occurring in the stability criteria has a familiar look, being the reproducing kernel $k(x,t)$ for F^π (p. 59) with one of its arguments evaluated at λ_n. Indeed the stability theory for nonharmonic Fourier functions brings out an interesting connection with the reproducing kernel theory for F^π. The kernel $k(x,t)$ can be obtained from the calculation of

$$\int_{-\pi}^{\pi} e^{ixw} e^{-itw}\, dw,$$

that is, the $L^2(-\pi,\pi)$ inner product of the Fourier kernel with itself; this calculation was necessarily involved in calculating the norm of the difference $e^{inx} - e^{i\lambda_n x}$. We have observed that $(k(x,n))$ is a CON sequence in F^π, generated by taking Fourier transforms of members of the CON sequence $((2\pi)^{-\frac{1}{2}} e^{inx})$ in $L^2(-\pi,\pi)$, and similarly we may observe that by the isometric character of the Fourier transform, any criterion guaranteeing that $(e^{i\lambda_n x})$ is a basis for $L^2(-\pi, \pi)$ is also a criterion guaranteeing that

$$\left(\frac{\sin \pi(x - \lambda_n)}{\pi(x - \lambda_n)}\right)$$

is a basis for F^π. These remarks generalise the remarks of § 2.6.3 to some extent; for an application to a problem in electrical engineering, see Higgins (1976).

A more general situation, of which we have just described a special case, is that of a subspace M of $L^2(\mathbb{R})$ generated by an isometric integral transform with kernel $K(x, t)$ operating on those functions of $L^2(\mathbb{R})$ with the finite interval $[a, b]$ as compact support. If the kernel has the property that, for some sequence (t_n), $(K(x, t_n))$ is a CON sequence in $L^2(a, b)$, then M necessarily has a reproducing kernel given by the inner product of K with itself over (a, b), and there is an interpolation theory in M analogous to the cardinal interpolation series in F^π. A further example of this is obtained by taking the Hankel transform and the associated Fourier–Bessel functions on $(0, 1)$ (see Higgins, 1972).

The literature on sets of non-harmonic Fourier functions is large and still growing; here we can only offer a small selection. A further example will be found in the next section.

3.2 A complex variable method

We have already met an application of complex function theory; this was when the identity theorem was used in the proof of the completeness theorem for polynomials (p. 31). In this section we shall use further powerful results from the theory of entire (integral) functions of exponential type, i.e. functions $f(z)$ regular in the whole plane, or in a sector of the plane, and for which there exist positive constants c and τ such that $|f(z)| < c e^{\tau |z|}$ as $z \to \infty$. In outline the method runs as follows.

Let a sequence $(\phi_n(x))$ to be tested for completeness, on L^p say, be such that for every n, $\phi_n(x) = f(\lambda_n x)$ for some real number λ_n. Then for $g \in L^p$ we shall want the assumption

$$\int_a^b f(\lambda_n x)\, g(x)\, dx = 0 \quad \text{for every } n$$

to imply that g is null. Suppose $f(z)$ is of exponential type, then by Hölder's inequality so also is

$$F(z) = \int_a^b f(zx)\, g(x)\, dx,$$

and $F(\lambda_n) = 0$ for every n. Now there are theorems which assert that an entire function $F(z)$ of exponential type is identically zero if $F(\lambda_n) = 0$ for certain sequences of points (λ_n), and if one of these theorems can be applied, the completeness of $(f(\lambda_n x))$ will have been demonstrated if we can show that $F(z) \equiv 0$ implies that g is null.

Typical of the required type is the following

THEOREM *Let $f(z)$ be regular and of exponential type for* $\operatorname{Re} z \geqslant \rho > 0$, *with zeros at the real points* $\lambda_1, \lambda_2, \dots$. *Let $n(x)$ denote the number of zeros of f not exceeding x and r and s be two real valued functions such that*

(i) $n(x) \geqslant x + r(x)$.

(ii) $\log |f(iy)f(-iy)| \leqslant 2\pi\{|y| + s(|y|)\}$.

(iii) $\overline{\lim\limits_{R \to 0}} \int_\rho^R \{r(y) - s(y)\}\, y^{-2}\, dy = +\infty$.

Then $f(z) \equiv 0$.

Proof The proof proceeds from Carleman's theorem, well known from elementary complex variable theory (see e.g. Titchmarsh (1939) p. 130) which we shall apply to $f(z)$ in the form

$$\sum_{\lambda_n \leqslant R} \left\{ \frac{1}{\lambda_n} - \frac{\lambda_n}{R^2} \right\} \leqslant \frac{1}{\pi R} \int_{-\pi/2}^{\pi/2} \log |f(Re^{i\theta})| \cos\theta\, d\theta$$

$$+ \frac{1}{2\pi} \int_\rho^R \left\{ \frac{1}{y^2} - \frac{1}{R^2} \right\} \log |f(iy)f(-iy)|\, dy + O(1) \quad (R \to \infty).$$

Since f is of exponential type, the first integral on the right-hand side is $O(1)$ as $R \to \infty$; using hypothesis (ii) the second integral does not exceed

$$\log R + \int_\rho^R s(y) \left\{ \frac{1}{y^2} - \frac{1}{R^2} \right\} dy + O(1)$$

so that the right-hand side does not exceed

$$\log R + \int_\rho^R s(y) \left\{ \frac{1}{y^2} - \frac{1}{R^2} \right\} dy + O(1).$$

Let us write the left-hand side as a Stieltjes integral, integrate by parts and then use hypothesis (i); these calculations yield

$$\begin{aligned}
\text{LHS} &= \int_\rho^R \left\{ \frac{1}{x} - \frac{x}{R^2} \right\} dn(x) \\
&= \int_\rho^R n(x) \left\{ \frac{1}{x^2} + \frac{1}{R^2} \right\} dx \\
&\geqslant \int_\rho^R (x + r(x)) \left\{ \frac{1}{x^2} + \frac{1}{R^2} \right\} dx \\
&= \int_\rho^R r(x) \left\{ \frac{1}{x^2} + \frac{1}{R^2} \right\} dx + \log R + O(1).
\end{aligned}$$

We may now write the original inequality as

$$\int_\rho^R r(x) \left\{ \frac{1}{x^2} + \frac{1}{R^2} \right\} dx \leqslant \int_\rho^R s(y) \left\{ \frac{1}{y^2} - \frac{1}{R^2} \right\} dy + O(1),$$

or

$$\int_\rho^R \{r(y) - s(y)\} \frac{1}{y^2} dy \leqslant O(1).$$

This contradicts hypothesis (iii) unless $f \equiv 0$, so the proof is complete.

EXAMPLE   The non-harmonic Fourier functions

$$\{e^{i\lambda_n x} : 0 < \lambda_1 < \lambda_2 < \ldots\}$$

form a complete sequence on $L(-a, a)$, and so on $L^p(-a, a)$, $p \geqslant 1$, if $\lim n/\lambda_n > a/\pi$, $a < \pi$.

Proof   Set $F(z) = \displaystyle\int_{-a}^a e^{izt} g(t) \, dt$, $g \in L(-a, a)$.

Then $|F(\pm iy)| = \left| \displaystyle\int_{-a}^a e^{\pm yt} g(t) \, dt \right|$

$$\leqslant e^{|y|a} \int_{-a}^a |g(t)| \, dt.$$

Let $F(\lambda_n) = 0$ $(n = 1, 2, ...)$. We apply the previous theorem to $\phi(z) = F(\pi z/a)$, which has zeros at $z = \lambda_n a/\pi$. Then

$$|\phi(\pm iy)| \leqslant e^{\pi|y|} \int_{-a}^{a} |g(t)| \, dt$$

therefore $\log |\phi(iy) \phi(-iy)| \leqslant 2\pi(|y| + \text{constant})$. Now if

$$\lambda_n a/\pi < x < \lambda_{n+1} a/\pi$$

then $n(x)/x > n\pi/\lambda_{n+1} a,$

therefore

$$\underline{\lim} \frac{n(x)}{x} \geqslant \frac{\pi}{a} \underline{\lim} \left\{ \frac{n+1}{\lambda_{n+1}} - \frac{1}{\lambda_{n+1}} \right\} > \frac{\pi}{a} \frac{a}{\pi} = 1,$$

therefore $\underline{\lim} \, n(x)/x \geqslant 1 + \alpha$, for some $\alpha > 0$. Also, for every $\epsilon > 0$, we have $n(x)/x > 1 + \alpha - \epsilon$, for sufficiently large x. We may choose $\epsilon < \alpha$, then $n(x) > x + (\alpha - \epsilon)x$ for all sufficiently large x. This condition is sufficient for the application of the previous theorem with $r(x) = (\alpha - \epsilon)x$; we also take $s = \text{constant}$. Hence

$$\overline{\lim_{R \to \infty}} \int_{1}^{R} \{r(y) - s(y)\} y^{-2} \, dy$$

$$= \overline{\lim_{R \to \infty}} \int_{1}^{R} (\alpha - \epsilon) y^{-1} \, dy + \text{constant}$$

$$= +\infty.$$

Therefore ϕ, and hence F, is identically zero. From the L^1 theory of Fourier transforms (e.g. Titchmarsh (1937) p. 164), g is null and the proof is complete.

Let us remark that if in the above example we take $\lambda_n = n$, then $\{e^{inx} : n > 0\}$ forms a complete sequence on $L^p(-a, a)$ if $a < \pi$. It is interesting to compare this with the standard result that $\{e^{inx} : n = 0, \pm 1, \pm 2, ...\}$ forms a complete sequence on $L^p(-\pi, \pi)$, where of course we cannot remove any member of the set without destroying the completeness; but if the interval $(-\pi, \pi)$ is shrunk by however small an amount to $(-a, a)$ we can have a complete sequence with only 'half as many' members as in the standard case! On the other hand, the theorem from which we deduced the above example is not powerful enough to yield the standard result itself. Consequently we give below a stronger theorem from which deeper completeness results can be obtained, as in the next section.

THEOREM (Boas and Pollard, 1947)   *Let*

(i) (λ_n) *be a sequence of reals satisfying*

$$0 < \lambda_n \leqslant n + \tfrac{1}{2}\alpha - \tfrac{1}{2}, n = 1, 2, \ldots;$$

(ii) $H(z)$ *be an entire function such that* $H(\pm\lambda_n) = 0$ *for every* n;

(iii) $|H(x+iy)| \leqslant |y|^{-\gamma} \displaystyle\int_0^\pi h(t)\, s(t)\, e^{t|y|}\, dt$, *where* $h \in L^p(0, \pi)$, $1 \leqslant p < \infty$, *and either*

(a) $s(t) \equiv 1$, *in which case we must have* $\alpha \leqslant \gamma + 1 - 1/p$, *or*

(b) $s(t) = (\sin t)^{1-1/p}$, *in which case we must have* $\alpha \leqslant \gamma + 2 - 2/p$.

Then $H(z) \equiv 0$.

As with the previous theorem the proof, which will be omitted, depends on one of the theorems of elementary complex variable theory, this time on that of Jensen.

3.3   The non-orthogonal Fourier–Bessel and Legendre functions

The sequence of trigonometrical functions (e^{inx}) is not the only orthogonal sequence for which the indexing parameter n can be perturbed to yield non-orthogonal sequences which retain basis and completeness properties. In this section we shall consider sequences of perturbed Fourier–Bessel functions $x^{\frac{1}{2}}J_\nu(\lambda_n x)$ for sequences (λ_n) other than the zeros of J_ν, and perturbed Legendre functions $P_{\lambda_n}(x)$ for points (λ_n) other than the integers. We rely heavily on the complex variable method of the previous section.

EXAMPLE (Boas and Pollard, 1947) *If* $\nu > -\frac{1}{2}$, *the set* $\{x^{\frac{1}{2}} J_\nu(\lambda_n x)\}$ *forms a complete (hence total) sequence in* $L^p(0,1)$, $1 \leqslant p < \infty$, *if for all sufficiently large n we have*

$$0 < \lambda_n \leqslant \pi\left(n + \tfrac{1}{4} + \frac{\nu}{2} - \frac{1}{2p}\right)$$

$$= \beta_n + \frac{1}{2q}$$

where $\beta_n = \pi(n + \nu/2 - \frac{1}{4})$.

Before proving this result, let us observe that we can recover from it a completeness result for the ordinary Fourier–Bessel functions if $\nu \geqslant \frac{1}{2}$. To see this we use the asymptotic formula (Watson (1922) p. 506)

$$j_{n\nu} = \beta_n - \frac{4\nu^2 - 1}{8\beta_n}\left\{1 + \sum_{r=1}^{\infty} Q_r \beta_n^{-2r}\right\}$$

where Q_r are constants depending only on ν. Thus for sufficiently large n we shall have

$$0 \leqslant j_{n\nu} \leqslant \beta_n \quad \text{if} \quad \nu^2 \geqslant \tfrac{1}{4},$$
$$j_{n\nu} > \beta_n \quad \text{if} \quad \nu^2 < \tfrac{1}{4}.$$

Thus the statement of the example applies to the ordinary orthogonal Fourier–Bessel functions if $\nu^2 \geqslant \frac{1}{4}$ and does not apply if $\nu^2 < \frac{1}{4}$.

Proof of the example By an obvious change of variable, the integral representation of Appendix 2,9, becomes

$$x^{\frac{1}{2}} J_\nu(\lambda x) = C x^{\frac{1}{2} - \nu} \int_0^x (x^2 - u^2)^{\nu - \frac{1}{2}} \cos \lambda u \, du, \ \nu > -\tfrac{1}{2}, \qquad (4)$$

where C is a constant depending on ν and λ. For completeness we must show that, for $f \in L^p(0, 1,)$,

$$\int_0^1 f(x) \, x^{\frac{1}{2}} J_\nu(\lambda_n x) \, dx = 0 \text{ for every } n \qquad (5)$$

implies f null. By substitution from (4) and a formal interchange of the order of integration, (5) becomes

$$C \int_0^1 g(u) \cos \lambda_n u \, du = 0 \text{ for every } n,$$

where

$$g(u) = \int_u^1 \frac{x^{\frac{1}{2} - \nu} f(x) \, dx}{(x^2 - u^2)^{\frac{1}{2} - \nu}}.$$

We leave it to the reader to show that there is no loss of generality if we assume that this integral always exists and defines a function $g(u)$. The reader is also asked to justify the various interchanges in the order of integration used throughout the proof. That g is in fact null follows from the previous theorem, for if we set

$$H(z) = \int_u^1 g(u) \cos zu \, du$$

then $H(z\pi)$ is an even entire function of exponential type with zeros at λ_n/π, $n = 0, 1, \ldots$ Furthermore, because

$$|\cos zu\pi| \leqslant e^{|y|u\pi} \, (z = x + iy),$$

and $|t/(t^2 - u^2)|^{\frac{1}{2}-\nu} \leqslant [2(t-u)]^{\nu-\frac{1}{2}} \, (u \leqslant t \leqslant 1),$
we have

$$2^{\frac{1}{2}-\nu}| \, H((x+iy)\,\pi)| \leqslant \int_0^1 e^{|y|u\pi} \left\{ \int_u^1 |f(t)|(t-u)^{\nu-\frac{1}{2}} \, dt \right\} du$$

$$= \int_0^1 |f(t)| \left\{ \int_0^t e^{|y|u\pi}(t-u)^{\nu-\frac{1}{2}} \, du \right\} dt$$

$$= \int_0^1 e^{|y|t\pi} |f(t)| \left\{ \int_0^t e^{-|y|u\pi} u^{\nu-\frac{1}{2}} \, du \right\} dt$$

$$< \Gamma(\nu+\tfrac{1}{2}) \, \pi^{-\nu-\frac{3}{2}} y^{-\nu-\frac{1}{2}} \int_0^\pi e^{|y|t} |f(t/\pi)| dt,$$

using the gamma function integral (Appendix 2,1). In the theorem (p. 88), case (*a*), take $s(t) = 1$, $\gamma = \nu + \frac{1}{2}$,

$$h(t) = 2^{\nu-\frac{1}{2}}\pi^{\nu+\frac{3}{2}}\Gamma(\nu+\tfrac{1}{2})f(t/\pi) \in L^p(0, \pi).$$

Now since $\lambda_n/\pi \leqslant n + 1/4 + \nu/2 - 1/2p$ we can take

$$\alpha = \nu + 3/2 - 1/p,$$

in which case $\alpha = \gamma + 1 - 1/p$ and the theorem shows that $H(z,\pi)$, and hence $H(z)$, is identically zero. By the uniqueness of Fourier cosine transforms g is null.

We now show that this implies that f too is null, which will complete the proof.

We treat first the case $-\frac{1}{2} < \nu < \frac{1}{2}$; in this case

$$
\begin{aligned}
0 &= \int_y^1 \frac{u g(u)\, du}{(u^2 - y^2)^{\frac{1}{2}+\nu}} \\
&= \int_y^1 \frac{u}{(u^2 - y^2)^{\frac{1}{2}+\nu}} \left\{ \int_u^1 \frac{x^{\frac{1}{2}-\nu} f(x)}{(x^2 - u^2)^{\frac{1}{2}-\nu}}\, dx \right\} du \\
&= \int_y^1 x^{\frac{1}{2}-\nu} f(x) \left\{ \int_y^x \frac{u\, du}{(u^2 - y^2)^{\frac{1}{2}+\nu}(x^2 - u^2)^{\frac{1}{2}-\nu}} \right\} dx \\
&= \frac{1}{2} \int_y^1 x^{\frac{1}{2}-\nu} f(x) \left\{ \int_{y^2}^{x^2} \frac{dt}{(t - y^2)^{\frac{1}{2}+\nu}(x^2 - t)^{\frac{1}{2}-\nu}} \right\} dx \\
&= \frac{\pi}{2 \sin (\nu + \frac{1}{2}) \pi} \int_y^1 x^{\frac{1}{2}-\nu} f(x)\, dx,
\end{aligned}
$$

using the beta function integral (Appendix 2,4), and the required conclusion follows.

When $\nu \geqslant \frac{1}{2}$, we can employ differentiation under the integral to reduce the case to one similar to that just treated; in fact the required conclusion follows from consideration of $g^{(k)}(u)$, where $k = [\nu - \frac{1}{2}] + 1$, if $\nu + \frac{1}{2}$ is not an integer, and $k = \nu - \frac{1}{2}$, if $\nu + \frac{1}{2}$ is an integer.

Problem

3.6 Formulate and prove a similar result to the previous example in which J_ν is replaced by the Struve function $\mathbf{H}_\nu$ (Appendix 2,10).

EXAMPLE $\{P_{\lambda_n}(x): n = 1, 2, \ldots\}$ forms a complete sequence on $L^p(-1, 1)$, $1 < p \leqslant 2$, if $-\frac{1}{2} < \lambda_n \leqslant n - 1/r$, $1 \leqslant r < p$ (cf. Boas and Pollard (1947) p. 375).

This example includes the completeness of the sequence of Legendre polynomials $(P_n(x))$ for the given range of p, by taking $\lambda_n = n - 1$; this is allowable since $r \geqslant 1$. Certain other sequences (λ_n) also yield orthogonal sequences $(P_{\lambda_n}(x))$ (see Hille, 1918) and the example also includes the completeness of these.

Proof of the example   Set

$$H^*(z) = \int_{-1}^{1} P_z(t)f(t)\, dt,\ f\in L^p(-1,1), z = x+iy.$$

We must show that $H^*(\lambda_n) = 0$ $(n = 1, 2, ...)$ implies that f is null. Using the integral representation (Appendix 2,6) for P_z we have

$$\pi 2^{-\frac{1}{2}}H^*(z) = \int_{-1}^{1} f(t)\left\{\int_{0}^{\cos^{-1}t} \frac{\cos(z+\frac{1}{2})\theta\, d\theta}{(\cos\theta - t)^{\frac{1}{2}}}\right\} dt$$

$$= \int_{0}^{\pi}\left\{\int_{-1}^{\cos\theta} \frac{f(t)\, dt}{(\cos\theta - t)^{\frac{1}{2}}}\right\}\cos(z+\tfrac{1}{2})\theta\, d\theta.$$

Set
$$H(z) = \int_{0}^{\pi} G(\theta)\cos z\theta\, d\theta,$$

where
$$G(\theta) = \int_{-1}^{\cos\theta} \frac{f(t)\, dt}{(\cos\theta - t)^{\frac{1}{2}}}.$$

In the first part of the proof we invoke the previous theorem to show that $H(z) \equiv 0$; in the second part we find that this implies that G is null, and in the third part that G null implies f null. Thus the proof will be completed.

1st part $H(z)$ is an even entire function of exponential type with zeros at $\lambda_n + \frac{1}{2}$ by assumption. Further,

$$|H(x+iy)| \leqslant \int_{0}^{\pi} e^{\theta|y|}\left\{\int_{-1}^{\cos\theta} \frac{|f(t)|\, dt}{(\cos\theta - t)^{\frac{1}{2}}}\right\} d\theta$$

$$= \int_{-1}^{1} |f(t)|\left\{\int_{0}^{\cos^{-1}t} \frac{e^{\theta|y|}\, d\theta}{(\cos\theta - t)^{\frac{1}{2}}}\right\} dt$$

$$= \int_{-1}^{1} |f(t)|\left\{\int_{t}^{1} \frac{e^{|y|\cos^{-1}u}\, du}{(1-u^2)^{\frac{1}{2}}(u-t)^{\frac{1}{2}}}\right\} dt$$

$$\leqslant \int_{-1}^{1} |f(t)|e^{|y|\cos^{-1}t}\left\{\int_{t}^{1} \frac{du}{(1-u^2)^{\frac{1}{2}}(u-t)^{\frac{1}{2}}}\right\} dt.$$

Denote the inner integral by I. There are two cases. When $t \geqslant 0$ we have

$$I \leqslant \int_{t}^{1} \frac{du}{(1-u)^{\frac{1}{2}}(u-t)^{\frac{1}{2}}} = \pi,$$

taking into account the special integral

$$\int \frac{du}{(a+u)^{\frac{1}{2}}(1-u)^{\frac{1}{2}}} = 2\tan^{-1}\left\{\frac{a+u}{1-u}\right\}^{\frac{1}{2}}. \tag{6}$$

When $-1 < t < 0$ we have

$$I \leqslant \int_t^0 \frac{du}{(u+|t|)^{\frac{1}{2}}(1+u)^{\frac{1}{2}}} + \int_0^1 \frac{du}{(u+|t|)^{\frac{1}{2}}(1-u)^{\frac{1}{2}}}$$

$$= I_1 + I_2, \text{ say.}$$

Using (6) again we find that $I_2 < \pi$, whilst use of the special integral

$$\int \frac{du}{(a+u)^{\frac{1}{2}}(1+u)^{\frac{1}{2}}} = 2\log\left[(a+u)^{\frac{1}{2}} + (1+u)^{\frac{1}{2}}\right]$$

yields $I_1 = 2\log\left(1 + |t|^{\frac{1}{2}}\right) - \log\left(1 + t\right)$. In either case there is a constant A such that $I \leqslant A + |\log(1+t)|$. We may therefore write

$$|H(x+iy)| \leqslant \int_0^\pi |f(\cos u)|\, e^{|y|u} \left(|\log(1+\cos u)| + A\right) \sin u\, du.$$

For purposes of applying the previous theorem we write this integral as

$$\int_0^\pi |f(\cos u)|\, e^{|y|u} \left(|\log(1+\cos u)| + A\right) (\sin u)^{1/r} (\sin u)^{1-1/r}\, du,$$

where r is to be chosen. In fact we shall have

$$|f(\cos u)|(\log(1+\cos u) + A)(\sin u)^{1/r} \in L^r(0,\pi)$$

because, in the first place,

$$\int_0^\pi |f(\cos u)|^r |\log(1+\cos u)|^r \sin u\, du$$

$$= \int_{-1}^1 |f(t)|^r |\log(1+t)|^r\, dt$$

$$\leqslant \left\{ \int_{-1}^1 |f(t)|^p\, dt \right\}^{r/p} \left\{ \int_{-1}^1 |\log(1+t)|^{rp/(p-r)}\, dt \right\}^{1-r/p},$$

for $p/r > 1$, i.e. $r < p$, and the last integral exists (by, for example, comparison with the gamma function integral). In the second place, that part of the integrand containing A also lies in $L^r(0,\pi)$, and the required conclusion follows with r chosen so that $1 \leqslant r < p$.

We now apply the previous theorem (case (*b*), $\gamma = 0$) to show that $H(z)$, and hence $H^*(z)$, is identically zero, provided that $\alpha \leqslant 2 - 2/r$, i.e.

$$0 < \lambda_n + \tfrac{1}{2} \leqslant n + 1 - 1/r - \tfrac{1}{2},$$

or

$$-\tfrac{1}{2} < \lambda_n \leqslant n - 1/r, \ 1 \leqslant r < p.$$

This completes the first part.

2nd part From the first part, that function equal to 0 for $\theta > \pi$ and equal to $G(\theta)$ for $0 < \theta < \pi$ has a vanishing Fourier cosine transform. From the Fourier transform theory (see Titchmarsh (1937) p. 96) $G(\theta)$ will be null if it lies in L^s for some s such that $1 < s \leqslant 2$. Indeed we show that $G(\theta) \in L^s(0, \pi)$ for every $s < p/(2-p)$:

$$\int_0^\pi |G(\theta)|^s d\theta = \int_{-1}^1 \frac{1}{(1-u^2)^{\frac{1}{2}}} \left| \int_{-1}^u \frac{f(x)\,dx}{(u-x)^{\frac{1}{2}}} \right|^s du$$

$$\leqslant \left\{ \int_{-1}^1 \left| \int_{-1}^u \frac{f(x)\,dx}{(u-x)^{\frac{1}{2}}} \right|^\alpha du \right\}^{s/\alpha} \left(\int_{-1}^1 \frac{du}{(1-u^2)^{\alpha/2(\alpha-s)}} \right)^{(\alpha-s)/\alpha}$$

for any α such that $\alpha/s > 1$. Now in the first term

$$\int_{-1}^u \frac{f(x)\,dx}{(u-x)^{\frac{1}{2}}} \in L^\alpha(-1,1), \ \alpha \leqslant 2p/(2-p), \quad 1 < p < 2,$$

by a special result in the theory of the fractional derivative operator (Hardy *et al.* 1952), whilst in the second term the integral is finite if $\alpha/2(\alpha-s) < 1$; with $\alpha = 2p/(2-p)$ this is $s < p/(2-p)$. Thus we may always take an s for which $1 < s < 2$, if $p < 2$.

If $p = 2$ then $f \in L^p(-1,1)$ for $p < 2$, and the argument goes through as before.

3rd part Since $f \in L^p(-1,1)$ $(1 < p \leqslant 2)$, $f \in L(-1,1)$. We have

$$\pi \int_{-1}^x f(t)\,dt = \int_{-1}^x f(t) \left\{ \int_t^x \frac{du}{(x-u)^{\frac{1}{2}}(u-t)^{\frac{1}{2}}} \right\} dt$$

upon using (6) again. Then for every $x \in (-1,1)$,

$$\pi \int_{-1}^x f(t)\,dt = \int_{-1}^x \frac{1}{(x-u)^{\frac{1}{2}}} \left\{ \int_{-1}^u \frac{f(t)\,dt}{(u-t)^{\frac{1}{2}}} \right\} du$$

$$= \int_{\cos^{-1}x}^\pi \frac{G(\phi)\sin\phi\,d\phi}{(x-\cos\phi)^{\frac{1}{2}}} = 0.$$

This shows that f is null and our proof is complete.

3.4 Some theorems of Müntz and Szász

The idea of perturbing the indexing parameter of a known complete set was first applied to the set of powers $\{x^n\}$ over $(0, 1)$. We shall prove the classical Müntz' theorem, the proof of which provides a further example of geometrical arguments in Hilbert space, and quote some more general results of Szász (whose proofs are to be found in Paley and Wiener (1934)).

THEOREM (Müntz)   *The set $\{x^{\lambda_1}, x^{\lambda_2}, \dots: 1 \leqslant \lambda_1 < \lambda_2 < \dots\}$ forms a complete sequence in $L^2(0, 1)$ if and only if $\Sigma \lambda_n^{-1} = \infty$.*

Proof   Consider the closed linear manifold $M_k = [x^{\lambda_1}, \dots, x^{\lambda_k}]$. Let $d_{k,m}$ denote the minimum distance of x^m to M_k; then because of the totality of (x^m) in $L^2(0, 1)$ we obtain the totality of (x^{λ_m}) if and only if the condition $\lim_{k\to\infty} d_{k,m} = 0$ holds for every $m = 1, 2, \dots$

Let P_k denote orthogonal projection onto M_k, and suppose that

$$P_k x^m = \sum_{i=1}^{k} a_i x^{\lambda_i};$$

taking account of $x^m - P_k x^m \perp M_k$ we shall have

$$(x^m, x^{\lambda_j}) - \sum_{i=1}^{k} a_i(x^{\lambda_i}, x^{\lambda_j}) = 0, \tag{7}$$

and

$$\begin{aligned} d_{k,m}^2 &= \|x^m - P_k x^m\|^2 \\ &= (x^m - P_k x^m, x^m) \\ &= \|x^m\|^2 - \sum_{i=1}^{k} a_i(x^{\lambda_i}, x^m). \end{aligned}$$

We shall write this expression in a different form by introducing G, the Gram determinant of $\{x^{\lambda_1}, \dots x^{\lambda_k}\}$ (see Appendix 1,7) and G^*, the Gram determinant of $\{x^m, x^{\lambda_1}, \dots x^{\lambda_k}\}$. Then

$$\frac{G^*}{G} = \frac{1}{G}\{(x^m, x^m)G - (x^m, x^{\lambda_1})M_1 + \dots + (-1)^k(x^m, x^{\lambda_k})M_k\},$$

where M_i is the minor of (x^m, x^{λ_i}) in G^*. In fact $M_i = (-1)^{i-1}G^{(i)}$, where $G^{(i)}$ is obtained from G by replacing its ith column with (x^m, x^{λ_j}); therefore

$$\frac{G^*}{G} = \|x^m\|^2 - \sum_{i=1}^{k} (x^m, x^{\lambda_i})\frac{G^{(i)}}{G}$$

$$= d_{k,m}^2,$$

since $a_i = \dfrac{G^{(i)}}{G}$ from (7) and Cramer's rule.

Now G^* and G can be evaluated; for

$$(x^p, x^q) = \int_0^1 x^{p+q}\,dx = (p+q+1)^{-1},$$

and an $n \times n$ determinant such as $\Delta_n = \det\left(\dfrac{1}{\alpha_i + \beta_j}\right)$ can be evaluated by subtracting the last row from each of the others and removing common factors, then again subtracting the last column and removing common factors, to obtain the recurrence relation

$$\Delta_n = \frac{\displaystyle\prod_{i=1}^{n-1}(\alpha_n - \alpha_i)(\beta_n - \beta_i)}{\displaystyle\prod_{i=1}^{n}(\alpha_n + \beta_i)\prod_{i=1}^{n-1}(\alpha_i + \beta_n)}\,\Delta_{n-1}$$

from which Δ_k is evaluated. Such calculations yield

$$d_{k,m}^2 = \frac{\displaystyle\prod_{i=1}^{k}(\lambda_i - m)^2}{(2m+1)^2\displaystyle\prod_{i=1}^{k}(m + \lambda_i + 1)^2}.$$

Then our condition becomes

$$\lim_{k\to\infty}\sum_{i=1}^{k}\left\{\log\left(1 - \frac{m}{\lambda_i}\right) - \log\left(1 + \frac{m+1}{\lambda_i}\right)\right\} = -\infty \text{ for every } m.$$

Now $\Sigma \log(1 + u_i)$ and Σu_i converge or diverge together, and the result follows from this.

THEOREM (Szász) *Let (λ_n) be a sequence of complex numbers such that* Re $\lambda_n > -\frac{1}{2}$ $(n = 1, 2, \ldots)$. *Then (x^{λ_n}) is complete in $L^2(0, 1)$ if and only if*

$$\sum_{n=1}^{\infty} \frac{1 + 2\operatorname{Re}\lambda_n}{1 + |\lambda_n|^2} = \infty.$$

THEOREM (Szász) *Let (λ_n) be a sequence of complex numbers such that* Re $\lambda_n > 0$ $(n = 1, 2, \ldots)$. *Then $\left(\dfrac{1}{1 + \lambda_n^2 x}\right)$ forms a complete sequence in $L^2(0, \infty)$ if and only if*

$$\sum_{n=1}^{\infty} \frac{\operatorname{Re}\lambda_n}{1 + |\lambda_n|^2} = \infty. \tag{8}$$

THEOREM (Szász' form of Müntz' theorem) *Let (λ_n) be a sequence of complex numbers such that* Re $\lambda_n > 0$ $(n = 1, 2, \ldots)$. *Then (x^{λ_n}) is total in $C(0, 1)$ if (8) holds, and fails to be total if the series converges.*

Whilst it is our declared purpose to concentrate on the L^2 and L^p aspects of completeness and basis properties, the presence of the Banach space $C(0, 1)$ in the above theorems may serve as a reminder that there is a substantial body of results concerning the *pointwise* convergence of expansions in the sets we have been studying. A large amount of information can be traced through the references in the bibliography; e.g. see Sz.-Nagy (1965), Alexits (1961), Titchmarsh (1946) and the more recent Olevskii (1975).

4. *Differential and integral operators*

This chapter is intended to provide the reader with an appreciation of at least one area of applied mathematics in which complete sets have a natural place. To this end we shall give several examples of boundary value problems involving differential operators in which complete orthogonal sequences appear as eigenfunctions of the system; we shall also look very briefly at the analogous situation for integral operators. The presentation is intended to be rather more a collection of useful facts than a detailed introduction to the subject; much will be quoted but little proved. Adequate references to more detailed sources are given; for example, Sagan (1961) is a good introductory text.

If T is an operator mapping a Hilbert space H to itself, it may happen that there is an $f \in H$ with the particularly simple property of being mapped by T onto a scalar multiple of itself, i.e.

$$Tf = \lambda f. \tag{1}$$

The structure of T is very closely tied up with sets of fs and λs for which (1) holds, as revealed by the so-called *spectral theory* of operators. Here we shall be concerned with a fairly simple situation, that in which there is a collection $\{\lambda_n\}$ of isolated points on the real axis such that for every n

$$Tf_n = \lambda_n f_n \tag{2}$$

for a certain non-null f_n, or collection of f_n, corresponding to λ_n. Then $\{\lambda_n\}$ is called the *point spectrum* of T and the individual λ_ns the *eigenvalues* of T. An f_n for which (2) is true is called an *eigenvector* of T, or *eigenfunction* if H is a function space. There may be several f_n corresponding to the same λ_n; the collection of all such f_n is called the *eigenspace* E_n corresponding to λ_n.

The examples of this chapter will chiefly be of the type where T of (1) is a differential operator so that (1) is a differential equation, and this will be accompanied by a suitable set of boundary conditions. These so-called *boundary value problems* arise typically in problems where one of the standard partial differential equations of Physics, e.g. Laplace's equation (in three dimensions)

$$\nabla^2 \phi = 0 \quad \left(\nabla^2 \equiv \frac{\partial^2}{\partial x^2} + \frac{\partial^2}{\partial y^2} + \frac{\partial^2}{\partial z^2} \right),$$

the wave equation

$$\nabla^2 \phi = \frac{1}{c^2} \frac{\partial^2 \phi}{\partial t^2},$$

or the equation of heat conduction

$$\nabla^2 \phi = \frac{1}{a^2} \frac{\partial \phi}{\partial t},$$

is cast into an appropriate coordinate system and then solved by the very well-known method of separation of variables (not to be confused with the method of the same name in § 3.2). Ordinary differential equations in each of the separate variables result, the solutions of which represent certain fundamental 'modes', e.g. of vibrations of a mechanical system. This is why the word 'spectrum' (from the Latin *spectare*, to look) has been borrowed from Physics, where it is used to mean an image of those fundamental parts into which a source of radiation can be decomposed by a certain physical process, and arranged in sequence according to wavelength.

It is the 'finite energy' condition, familiar from the theory of vibrations and wave motion in general, which provides the main reason for taking an L^2 space as the natural Hilbert space upon which the operators of this chapter should be considered to act. The finite energy condition is readily illustrated by an example taken from signal theory. Suppose that a signal is to be passed through a transmitting device; a standard engineering procedure is to take the signal $f(t)$ as being proportional to a voltage amplitude. If E (volts) denotes e.m.f. and R (ohms) denotes resistance then a well-known formula shows that power (watts) = E^2/R. That is, the power (the rate of expenditure of energy) is proportional to $|f(t)|^2$. The total energy of the signal must be finite, so that

$$\int_{-\infty}^{\infty} |f(t)|^2 \, dt < \infty.$$

Thus a finite energy signal is just an element of $L^2(\mathbb{R})$.

It need not concern us that differential operators are not everywhere defined on L^2 spaces; their domains are *dense* in such spaces and this is usually all that is required from the point of view of the theory of operators.

We shall introduce certain theorems guaranteeing the completeness of sequences of eigenfunctions, and these can sometimes be used to demonstrate completeness of a sequence already known to consist of the eigenfunctions of a suitable system. This method is of rather limited practical use, owing in large measure to the difficulties associated with singular systems (see the remarks beginning § 4.2.) However, it does provide a convenient method for treating the Mathieu functions (p. 104): see also Dahlberg (1973) where the 'flat-clamped-plate modes' (Appendix 3,24) are treated.

4.1 Sturm–Liouville systems

If T is taken to be a differential operator, given by

$$(Ty)(x) = (p_0(x) D^n + \ldots + p_{n-1}(x) D + p_n(x)) y(x)$$

then $Ty = \lambda y$, accompanied by a suitable set of boundary conditions, is called an *eigenvalue problem*. The situation is illustrated by the following example, which we shall discuss in some detail in order to exhibit some of the favourable properties that such boundary value problems possess.

EXAMPLE $\quad \begin{cases} -y'' = \lambda y \\ y'(0) = y'(\pi) = 0. \end{cases}$

First, let us assume that, for two different values λ_m and λ_n of λ there are corresponding solutions y_m and y_n of the differential equation (eigenfunctions of the problem), i.e. we have

$$-y_m'' = \lambda_m y_m \quad \text{and} \quad -y_n'' = \lambda_n y_n.$$

If we multiply the first equation by y_n and the second by y_m and then subtract we obtain

$$(\lambda_n - \lambda_m)\, y_n y_m = y_n y_m'' - y_m y_n''.$$

The right-hand side of this equation is $(y_n y_m')' - (y_m y_n')'$ so that integration from 0 to π yields

$$(\lambda_n - \lambda_m)\int_0^\pi y_n(t)\, y_m(t)\, dt = [y_n(t)\, y_m'(t) - y_m(t)\, y_n'(t)]_0^\pi = 0$$

by the boundary conditions. Since $\lambda_n \neq \lambda_m$ we have shown that y_n is orthogonal to y_m in the $L^2(0,\pi)$ sense. The reader will appreciate that we have obtained this important orthogonality property of the eigenfunctions without actually solving the differential equation for them!

It is easy to see that all eigenvalues must be real, for if λ_n is a complex eigenvalue with y_n as eigenfunction, then $\bar{\lambda}_n$ is an eigenvalue with $\bar{y}_n$ as eigenfunction; but no non-null function can be orthogonal to its complex conjugate.

We now show that such eigenfunctions do indeed exist, by solving for them. There are three separate possibilities:

(i) $\lambda < 0$. The solution of the equation is

$$y = A e^{\sqrt{(-\lambda)}x} + B e^{-\sqrt{(-\lambda)}x}.$$

The boundary conditions give

$$A - B = 0$$

and

$$A e^{\sqrt{(-\lambda)}\pi} - B e^{-\sqrt{(-\lambda)}\pi} = 0.$$

For a non-trivial solution for A and B we must have

$$\begin{vmatrix} 1 & -1 \\ e^{\sqrt{(-\lambda)}\pi} & -e^{-\sqrt{(-\lambda)}\pi} \end{vmatrix} = 0$$

or

$$\cosh \sqrt{(-\lambda)}\,\pi = 0.$$

But this equation has no solution, hence $A = B = 0$, and the problem has only the trivial solution $y \equiv 0$.

(ii) $\lambda = 0$. The solution of the equation is $y = Ax + B$. The boundary conditions yield $A = 0$ and the problem has solution $y = \text{constant}$.

(iii) $\lambda > 0$. The solution is

$$y = A \sin \sqrt{(\lambda)}\, x + B \cos \sqrt{(\lambda)}\, x.$$

The boundary conditions yield $A = 0$, and

$$B \sin \sqrt{(\lambda)}\, \pi = 0.$$

To obtain a non-trivial solution we must have $B \neq 0$; then $\sqrt{\lambda} = n$, $n = \pm 1, \pm 2, \ldots$, i.e. $\lambda = n^2$, $n = 1, 2, \ldots$ The corresponding solutions are $y_n = B_n \cos nx$, $n = 1, 2, \ldots$

To summarise, we find that the original eigenvalue problem has *eigenvalues n^2 with corresponding eigenfunctions $B_n \cos nx$, $n = 0, 1, \ldots$, which form an orthogonal sequence in $L^2(0, \pi)$.*

These eigenfunctions are of course also complete in $L^2(0, \pi)$ as we already know, and although the orthogonality could have been verified independently as well, it is of interest to ask whether there is a class of eigenvalue problems having both orthogonality and completeness of the eigenfunctions as an intrinsic property. There do indeed exist such classes, the most important of which is the class defined as follows:

DEFINITION   A *second order Sturm–Liouville problem* is a boundary value problem of the form

$$\begin{cases} (p(x)\, y')' + q(x)\, y + \lambda w(x)\, y = 0 \\ c_1 y(a) + c_2 y'(a) = 0 \\ c_3 y(b) + c_4 y'(b) = 0 \end{cases}$$

on the finite interval $[a, b]$, in which p, p', q and w are continuous on $[a, b]$, p and w are strictly positive on $[a, b]$, c_1 and c_2 are scalars, not both zero, and similarly c_3 and c_4.

The differential equation is evidently an eigenvalue equation of the form (1). It is often convenient to transform this equation into *Liouville normal form*

$$\frac{d^2 u}{dt^2} + [\lambda - q^*(t)]\, u = 0 \quad \text{on} \quad [0, c]$$

by the substitutions $y(x)\, (p(x)\, w(x))^{\frac{1}{4}} = u(t)$ and

$$t = \int_a^x (w(\tau)/p(\tau))^{\frac{1}{2}}\, d\tau,$$

where
$$q^* = q w^{-1} + (pw)^{-\frac{1}{4}} \frac{d^2}{dt^2} ((pw)^{\frac{1}{4}}), \quad \text{and} \quad c = \int_a^b (w/p)^{\frac{1}{2}}.$$

The facts contained in the following theorem are to be found in the literature on Sturm–Liouville systems (e.g. Birkhoff and Rota, 1969) and their proofs will be omitted except for the interesting proof of the completeness of the eigenfunctions.

THEOREM (i) *The eigenvalues of a second order Sturm–Liouville problem are real, countably infinite and have no finite point of accumulation. To each eigenvalue there corresponds one and only one (up to a constant factor) eigenfunction; the collection of normalised eigenfunctions constitutes a CON sequence in* $L^2((a, b), w)$.

(ii) *The eigenvalues* (λ_n) *and eigenfunctions* (ϕ_n) *of a second order Sturm–Liouville problem on* $[a, b]$ *in which the differential equation is in Liouville normal form satisfy, respectively, the asymptotic formulae*

$$\lambda_n = n/(b-a) + O(1)/n$$

and $\quad \phi_n(x) = [2/(b-a)]^{\frac{1}{2}} \cos [n\pi(x-a)/(b-a)] + O(1)/n$

as $n \to \infty$.

These properties bring out the essential character of the Sturm–Liouville problem; it is seen that all the properties we found for our previous example carry over to the more general case, either directly or in asymptotic form. It is from the asymptotic form for the eigenfunctions that we can deduce their completeness:

Proof of the completeness of the eigenfunctions First we transform to Liouville normal form. The substitutions involved induce a mapping of $L^2((a,b), w)$ onto $L^2(0, c)$ which is an isometric isomorphism: for example norms are preserved since, if y of $L^2((a, b), w)$ is transformed into u, we have

$$\int_a^b |y(x)|^2 w(x)\, dx = \int_a^b |y(x)|^2 [p(x)\, w(x)]^{\frac{1}{2}} [w(x)/p(x)]^{\frac{1}{2}}\, dx$$

$$= \int_0^c |u(t)|^2\, dt.$$

We have seen that CON sequences correspond under an isometric isomorphism, so to complete the proof we must show that the eigenfunctions (ϕ_n) of the problem in Liouville normal

5

form are complete in $L^2(0, c)$. The asymptotic formula of the previous theorem together with the stability theorem A (p. 72) contain all the hard work and it merely remains to note that

$$\{\xi_n(t)\} = \{c^{-\frac{1}{2}}, (2/c)^{\frac{1}{2}}\cos(n\pi x/c): n = 1, 2, \ldots\}$$

forms a CON sequence in $L^2(0, c)$ (p. 37), and

$$\sum_{n=1}^{\infty} \|\phi_n - \xi_n\|^2 \leqslant \text{const} + (\text{const}) \sum_{n=1}^{\infty} n^{-2} < \infty.$$

EXAMPLE   Completeness of the periodic Mathieu functions. The differential equation

$$\frac{d^2y}{dx^2} + (\lambda - 2\theta\cos 2x)\,y = 0$$

is called Mathieu's equation. It arises when the two-dimensional wave equation (p. 99) is separated in elliptic coordinates; thus it arose in Mathieu's researches into the vibrations of an elliptic membrane. In such problems of mechanical vibrations it turns out that, for fixed θ, there are certain special values of λ for which the equation has periodic solutions; it is with these special solutions that we are concerned in this example. The equation occurs in many other physical problems as well, such as orbit problems in astronomy, electrical circuits with varying resistance, and vibrations of a string with varying tension. Solutions of the equation for various values of the parameters λ and θ are called Mathieu functions (see McLachlan (1947) for a comprehensive account of both theory and applications).

For a fixed value of θ the equation is evidently of the Sturm–Liouville type. In fact if $\theta = 0$, it is just the familiar oscillation equation $y'' + \lambda y = 0$, the sine and cosine solutions of which can be thought of as special cases of the periodic Mathieu functions, with which they have several points of analogy which we now quote:

For certain values $\{\lambda_n\}$ of λ (these values depend on θ of course; when $\theta = 0$, $\lambda_n = n^2$), Mathieu's equation has solutions denoted by

$$ce_0(x, \theta), ce_n(x, \theta), se_n(x, \theta), \quad n = 1, 2, \ldots,$$

such that

(i) they reduce to $1, \cos nx, \sin nx$

respectively, when $\theta = 0$;

(ii) they are periodic with period 2π;

(iii) the set $\{ce_0,\ ce_n,\ se_n\}$ forms an orthogonal sequence in $L^2(-\pi, \pi)$.

One naturally conjectures that this sequence is also complete in $L^2(-\pi, \pi)$; this is indeed the case and will follow readily from the Sturm–Liouville theory. We need to quote one further fact from the theory of Mathieu's equation (see e.g. Erdélyi *et al.* (1953) p. 112), namely that the functions $\{se_n(x, \theta) : n = 1, 2, \ldots\}$ are determined up to a constant factor as solutions of the boundary value problem

$$\begin{cases} y'' + (\lambda - 2\theta \cos 2x)\, y = 0 \\ y(0) = y(\pi) = 0, \end{cases}$$

and likewise $\{ce_n : n = 0, 1, \ldots\}$ from

$$\begin{cases} y'' + (\lambda - 2\theta \cos 2x)\, y = 0 \\ y'(0) = y'(\pi) = 0. \end{cases}$$

Note that ce_n and se_n could have been generated by imposing *periodic* boundary conditions on $(-\pi, \pi)$, i.e. $y(-\pi) = y(\pi)$ and $y'(-\pi) = y'(\pi)$, but these are not of the Sturm–Liouville type and the theory would not have been applicable. The complete orthogonal character of Sturm–Liouville eigenfunctions now shows that both $\{se_n : n = 1, 2, \ldots\}$ and $\{ce_n : n = 0, 1, \ldots\}$ form complete orthogonal sequences in $L^2(0, \pi)$. Then by obvious extensions to $(-\pi, \pi)$ we have (se_n) complete orthogonal in $L_o^2(-\pi, \pi)$ (the odd functions of $L^2(-\pi, \pi)$), and (ce_n) complete orthogonal in $L_e^2(-\pi, \pi)$ (the even functions of $L^2(-\pi, \pi)$). Now from $L^2 = L_o^2 \oplus L_e^2$ we obtain the required result: *The periodic Mathieu functions* $\{ce_0,\ ce_n,\ se_n\}$ *form a complete orthogonal sequence in* $L^2(-\pi, \pi)$.

Problems

4.1 Show that the following are Sturm–Liouville systems with the eigenvalues λ_n and eigenfunctions ϕ_n as shown. Also deduce the series expansions given for each by use of Vitali's criterion:

(i) $\begin{cases} y'' + \lambda y = 0 \\ y(0) = y'(\pi) = 0 \end{cases}$ $\begin{aligned} \lambda_n &= (n - \tfrac{1}{2})^2, n = 1, 2, \dots \\ \phi_n &= A_n \sin(n - \tfrac{1}{2})x. \end{aligned}$

$$\sum_{n=1}^{\infty} \left\{ \frac{1 - \cos(n - \tfrac{1}{2})x}{n - \tfrac{1}{2}} \right\}^2 = \pi x/2, x \in (0, \pi).$$

(ii) $\begin{cases} y'' + \lambda y = 0 \\ y'(0) = y(\pi) = 0 \end{cases}$ $\begin{aligned} \lambda_n &= (n - \tfrac{1}{2})^2, n = 1, 2, \dots \\ \phi &= A_n \cos(n - \tfrac{1}{2})x. \end{aligned}$

$$\sum_{n=1}^{\infty} \left\{ \frac{\sin^2(n - \tfrac{1}{2})x}{n - \tfrac{1}{2}} \right\}^2 = \pi x/2, x \in (0, \pi).$$

(iii) $\begin{cases} y'' + \lambda y = 0 \\ y(0) + y'(0) = 0 \\ y(\pi) + y'(\pi) = 0 \end{cases}$ $\begin{aligned} \lambda_0 &= -1, \text{ and } \lambda_n = n^2, n = 1, 2, \dots \\ \phi_0 &= e^{-x}, \text{ and } \phi_n = \sin nx - n \cos nx. \end{aligned}$

$$\frac{(1 - e^{-x})^2}{1 - e^{-2\pi}} + \frac{1}{\pi} \sum_{n=1}^{\infty} \left(\frac{\cos nx}{n} + \sin nx - \frac{1}{n} \right)^2 \Big/ (n^2 + 1) = \frac{x}{2}.$$

$$\sum_{n=1}^{\infty} \frac{1}{(2n^2 - 2n + 1)(2n - 1)^2} = \frac{\pi^2}{4} - \frac{\pi}{2} \frac{1 - e^{-\pi}}{1 + e^{-\pi}}.$$

4.2 Show that the following are Sturm–Liouville problems, and find their eigenvalues and eigenfunctions:

(i) $\begin{cases} y'' + 2y' + (1 + \lambda)y = 0 \\ y(0) = y'(a) = 0. \end{cases}$

(ii) $\begin{cases} x^2 y'' + xy' - \lambda y = 0 \\ y(1) = y(e) - ey'(e) = 0. \end{cases}$

(iii) $\begin{cases} (x^2 y')' + \dfrac{\lambda y}{x^2} = 0 \\ y(1) = y(2) = 0. \end{cases}$

(iv) $\begin{cases} x^2 y'' + \lambda y = 0 \\ y(1) = y(e^\pi) = 0. \end{cases}$

4.1.1 Generalisation of the Sturm–Liouville problem

We mention briefly that the scope of the Sturm–Liouville problem can be considerably widened in both the order of the equation and the nature of the boundary conditions, whilst still retaining the 'regular' character of the original problem, provided that certain conditions of 'self-adjointness' are satisfied.

Together with the differential operator

$$T = p_0 D^n + \dots + p_n$$

we have its *formal adjoint*, given by

$$T^*y = (-1)^n (\overline{p}_0 y)^{(n)} + (-1)^{n-1} (\overline{p}_1 y)^{(n-1)} + \dots + \overline{p}_n y.$$

If T coincides with its formal adjoint it is called *self-adjoint*.

Let there be associated with T the n *homogeneous boundary conditions*

$$\mathcal{B}_j y = \sum_{k=1}^{n} M_{jk} y^{(k-1)}(a) + N_{jk} y^{(k-1)}(b) = 0 \quad (j = 1, 2, \dots, n)$$

with matrices of coefficients $M = (M_{jk})$ and $N = (N_{jk})$.

We define $B(t)$ as being the matrix of coefficients $(B_{jk}(t))$ obtained when the formal expression

$$[uv](t) = \sum_{m=1}^{n} \sum_{j+k=m-1} (-1)^j u^{(k)}(t) (p_{n-m} \overline{v})^{(j)}(t)$$

is written

$$\sum_{j,k=1}^{n} B_{jk}(t) u^{(k-1)}(t) v^{(j-1)}(t);$$

then B is such that

$$B_{jk}(t) = \begin{cases} 0, & j+k > n+1 \\ (-1)^{j-1} p_0(t), & j+k = n+1, \end{cases}$$

and $\det B(t) = [p_0(t)]^n$ so that B is non-singular.

DEFINITION The system

$$\begin{cases} Ty = \lambda y & (p_j \in C^{n-j}; j = 0, 1, \dots n; p_0 \neq 0 \text{ in } [a, b]) \\ \mathcal{B}_j y = 0 & (j = 1, \dots, n) \end{cases}$$

is called a *regular self-adjoint boundary value problem* if T is self-adjoint, and the $\mathscr{B}_j$ satisfy the *self-adjointness condition*

$$MB^{-1}(a)\,M' = NB^{-1}(b)\,N',$$

where ′ denotes the conjugate transpose.

Whilst not all of the Sturm–Liouville features carry over to this more general case, two of the most important ones for our purposes do so, namely the orthogonality of eigenfunctions corresponding to distinct eigenvalues and the completeness of the sequence of eigenfunctions. The eigenspaces may not all be one-dimensional, however, but we can always orthogonalise a set of linearly independent eigenfunctions spanning a multi-dimensional eigenspace, and indeed we may state: *a regular self-adjoint boundary value problem possesses a set of eigenfunctions forming a CON sequence in* $L^2(a,b)$. There are generalisations of this result, for example, to the case of $L^2((a,b),w)$, and even to non-self-adjoint problems in which the eigenfunctions form a Riesz basis for $L^2(a,b)$, with the eigenfunctions of the adjoint problem as biorthogonal set. For further information see Coddington and Levinson (1955) and Naimark (1968); the latter reference contains a synthesis of the pioneering work of G. D. Birkhoff, Tamarkin, Stone *et al.* dating from the early years of this century.

EXAMPLE We generate the familiar trigonometrical set $\{1, \cos nx, \sin nx\}$ on $(-\pi, \pi)$ from the boundary value problem

$$\begin{cases} y'' + \lambda y = 0 \\ y(-\pi) = y(\pi) \\ y'(-\pi) = y'(\pi). \end{cases}$$

Here we have imposed periodic boundary conditions which, as we have already noted are not of the Sturm–Liouville type; the consequence is that we do not retain the Sturm–Liouville one-dimensional eigenspace feature. The calculations are very similar to those of the first example of §4.1, and we merely quote the results:

(i) $\lambda < 0$: trivial solution $y \equiv 0$ only.

(ii) $\lambda = 0$: solution $y = $ constant.

(iii) $\lambda > 0$: general solution $y = A \sin \sqrt{(\lambda)}x + B \cos \sqrt{(\lambda)}x$, subject to the condition $\sin \sqrt{(\lambda)}\,\pi = 0$. Therefore when $\lambda = n^2$, $n = 1, 2, ...$, there are two linearly independent eigenfunctions, so that in this case each eigenspace is of dimension 2.

This problem, although not Sturm–Liouville, is self-adjoint according to the above definition. For the equation is obviously self-adjoint, and as for the boundary conditions we have

$$M = \begin{bmatrix} 1 & 0 \\ 0 & 1 \end{bmatrix} \qquad N = \begin{bmatrix} -1 & 0 \\ 0 & -1 \end{bmatrix} \qquad B = \begin{bmatrix} 0 & -1 \\ 1 & 0 \end{bmatrix}$$

and the self-adjointness condition is readily verified from these. We have another proof of the completeness in $L^2(-\pi, \pi)$ of the trigonometrical functions.

4.2 Singular boundary value problems

Our examples of boundary value problems up to this point have been 'regular', that is they have involved a differential equation on a finite interval with no singularities, and with the exception of the Mathieu functions, there has been a marked lack of interesting sets of eigenfunctions. It frequently happens that boundary value problems arising in practice involve a differential equation with singularities, or on a basic interval which is infinite in extent; with suitable boundary conditions (which may not be expressible in convenient formulae as in the regular case) such problems are called *singular boundary value problems*. Needless to say the convenient properties of the regular problem cannot always be expected to carry over to these singular cases, or not without a struggle if they do, and this is a pity because the singular cases constitute a class 'which seems to include all the most interesting examples', as E. C. Titchmarsh (1946) points out in his introduction. For example the spectrum of a singular problem may consist of a continuum rather than a set of discrete points; in such cases the eigenfunction expansion may be replaced with an integral representation.

If a countable set of eigenfunctions does exist, its orthogonality may often be established directly from the equation as in our previous examples, provided that the boundary conditions make the integrated term vanish. Completeness of the eigenfunctions is not so readily obtained however; it may sometimes be established by regarding the singular problem as a limit of a sequence of regular problems (see Coddington and Levinson (1955) p. 224). If it is required to demonstrate completeness of a sequence of eigenfunctions arising from a singular boundary value problem it may be just as easy to proceed independently of the theory of differential equations altogether, especially if the set is orthogonal, possibly by using one of the methods of the previous chapters.

EXAMPLES of singular boundary value problems with complete sets of eigenfunctions:

(1) Three standard cases arise from separating Laplace's equation $\nabla^2 \phi = 0$ in (i) cartesian, (ii) cylindrical polar, and (iii) spherical polar coordinates. The standard oscillation equation

$y'' + \lambda y = 0$ arises in all three cases.

In case (i) no other equation arises.

In case (ii) there arises the Bessel equation

$$y'' + \frac{1}{x}y' + \left(\lambda - \frac{\nu^2}{x^2}\right)y = 0$$

with a singularity at $x = 0$. If we put this into Liouville normal form and attach a natural boundary condition we get the singular boundary value problem

$$\begin{cases} y'' + \left(\lambda - \frac{4\nu^2 - 1}{4x^2}\right)y = 0 \\ y \text{ bounded at } x = 0 \end{cases}$$

on $[0, 1]$, with eigenvalues $\lambda_n = j_n^2$, $n = 1, 2, \ldots$ (where $J_\nu(j_n) = 0$, $j_n \neq 0$) and the complete orthogonal Fourier–Bessel functions $\{x^{\frac{1}{2}} J_\nu(j_n x)\}$ as corresponding eigenfunctions.

In case (iii) there arises the associated Legendre equation with singularities at 1 and -1. This yields the singular problem

$$\begin{cases} (1 - x^2)y'' - 2xy' + (\lambda - m^2(1 - x^2)^{-1})y = 0 \\ y \text{ bounded at } x = \pm 1 \end{cases}$$

on $[+1, 1]$, with eigenvalues $\lambda_n = n(n+1)$, $n = 0, 1, \dots$, and the complete orthogonal associated Legendre functions $\{P_n^m(x)\}$ as corresponding eigenfunctions.

(2) The differential equation

$$y'' + (\lambda - \phi(x))\, y = 0,$$

in which ϕ is called a *potential function*, is the one-dimensional Schrödinger equation which occurs in quantum theory. The equation is already in Liouville normal form, and it will give rise to a singular problem if the basic interval is taken to be infinite in extent, for example the whole real axis. Such problems may or may not possess a countable set of eigenfunctions, owing to the peculiar nature of the spectrum. We illustrate with two examples:

(I) zero potential:

$$\begin{cases} y'' + \lambda y = 0 \quad \text{on } \mathbb{R} \\ y \text{ bounded on } \mathbb{R}. \end{cases}$$

As in previous examples involving this equation there are three cases:

(i) $\lambda < 0$. Solutions of the equation are $\sinh \sqrt{(\lambda)}\, x$ and $\cosh \sqrt{(\lambda)}\, x$ neither of which are bounded on $\mathbb{R}$.

(ii) $\lambda = 0$. The only bounded solution is $y = \text{constant}$.

(iii) $\lambda > 0$. Bounded solutions are $\sin \sqrt{(\lambda)}\, x$ and $\cos \sqrt{(\lambda)}\, x$ for any positive real λ (these are not in $L^2(\mathbb{R})$ of course).

Thus the spectrum consists of the continuum $\lambda \geqslant 0$. The usual eigenfunction expansion is replaced by the Fourier integral representation.

(II) potential proportional to x^2: the singular problem

$$\begin{cases} y'' + (\lambda + x^2)\, y = 0 \\ y \to 0 \text{ as } x \to \pm \infty \end{cases}$$

is that of the so-called 'harmonic oscillator' on $\mathbb{R}$. The eigenvalues are $\lambda_n = 2n + 1$, $n = 0, 1, \dots$, with corresponding eigenfunctions $\{\exp(-x^2/2)\, H_n(x)\}$, the complete orthogonal Hermite functions.

In this last example the spectrum was again discrete; however Schrödinger's equation can lead to spectra with an even more peculiar character than that of example (1) above. For example there are cases of spectra consisting of a discrete part with a finite point of accumulation, and a continuous part.

Many other similar examples are to be found treated in some detail by Vogel (1953).

Problems

Show that the following differential equations are self-adjoint, and find boundary conditions yielding the singular boundary value problems with the eigenvalues and corresponding eigenfunctions indicated (see Meux, 1966).

4.3 $((1-x)^2(1+x)^2 y'')'' + (2(x^2-1)y')' + \lambda y = 0$ on $[-1, \ 1]$;
$\lambda_n = -n^2(n+1)^2$; $\{P_n(x)\}$, $n = 0, 1, \ldots$

4.4 $(x^2 e^{-x} y'')'' - (x e^{-x} y')' - \lambda e^{-x} y = 0$ on $\mathbb{R}^+$; $\lambda_n = -n^2$;
$\{e^{-x} L_n(x)\}$, $n = 0, 1, \ldots$

4.5 $(\exp(-x^2/2)y'')'' - (\exp(-x^2/2) \ y')' - \lambda \exp(-x^2/2) \ y = 0$
on $\mathbb{R}$; $\lambda_n = -n^2$; $\{\exp(-x^2/2) \ H_n(x)\}$, $n = 0, 1, \ldots$

4.3 Integral operators

The purpose of this section is to quote some facts from the Hilbert–Schmidt theory of Fredholm integral equations in order to illustrate how complete sets can arise as eigenfunctions of certain integral operators. This material is available in many standard texts on integral equations, see e.g. Kanwal (1971) and Lovitt (1950).

If the operator T of our general eigenvalue equation $Tf = \lambda f$ is taken to be the integral operator

$$If = (If)(x) = \int_a^b K(x,t) f(t) \, dt$$

then the equation becomes $If = \lambda f$. From the point of view of the theory of integral equations it is usually more convenient to replace λ with its reciprocal and write

$$\lambda If = f. \tag{3}$$

The definition of I is purely formal; if it is to define an operator on, say, $L^2(a,b)$ then K must be such that $If \in L^2(a,b)$ for $f \in L^2(a,b)$. This will certainly be the case if K is continuous in the finite square $S = \{a \leqslant x \leqslant b, \ a \leqslant t \leqslant b\}$. Actually it is sufficient that K be an $\mathscr{L}^2$ *kernel*, viz.

$$\iint_S |K|^2 < \infty.$$

$K(x,t)$ is called the *kernel* of the operator, and (3) is called a *homogeneous Fredholm integral equation*. There is a theory of eigenvalues and eigenfunctions (which are usually said to 'belong to K') of such equations analogous to that of the differential equations of the previous sections. For simplicity we restrict attention to real L^2 space in this section.

Perhaps the most obvious consequence of (3) is that if K is symmetric, that is $K(x,t) = K(t,x)$, then eigenfunctions corresponding to distinct eigenvalues are orthogonal. For if $f_n = \lambda_n \, If_n$ and $f_m = \lambda_m \, If_m$ then

$$\int_a^b f_n(x) f_m(x)\, dx = \lambda_n \int_a^b \int_a^b K(x,t) f_n(t) f_m(x)\, dt\, dx$$

$$= \lambda_m \int_a^b \int_a^b K(x,t) f_m(t) f_n(x)\, dt\, dx.$$

By interchange of x and t and use of the symmetry of K, we find that these two double integrals are equal, and by subtraction must be zero if $\lambda_n \neq \lambda_m$. This gives the required result.

We now quote three important theorems concerning the integral operator I. We shall need the following

DEFINITION The $\mathscr{L}^2$ symmetric kernel K is called *non-negative definite* if

$$\int_a^b \int_a^b K(x,t) f(x) f(y)\, dx\, dy \geqslant 0$$

for all $f \in L^2(a,b)$ $(f \neq 0)$. If K satisfies the above relationship with $\geqslant$ replaced by $>$ then it is called *positive definite*.

THEOREM *Let the kernel K be $\mathscr{L}^2$ and symmetric over the finite square S. Then it possesses at least one eigenvalue, all eigenvalues are real, constitute a countable set, and have no finite point of accumulation. The eigenfunctions corresponding to distinct eignevalues are orthogonal, and each eigenspace is of finite dimension.*

THEOREM (Hilbert–Schmidt) *Let $K(x, t)$ be a symmetric kernel, continuous on the finite square S, let g be piecewise continuous and*

$$f(x) = \int_a^b K(x,t)\, g(t)\, dt.$$

Then f can be expanded in the absolutely and uniformly convergent eigenfunction expansion

$$f(x) = \sum_{n=1}^{\infty} (f_n, f) f_n(x).$$

THEOREM (i) *A non-null symmetric $\mathscr{L}^2$ kernel is non-negative definite if and only if all its eigenvalues are positive.*

(ii) *A non-null symmetric $\mathscr{L}^2$ kernel is positive definite if and only if all its eigenvalues are positive and in addition the totality of eigenfunctions is complete in $L^2(a,b)$.*

Note that the eigenfunctions may *not* be complete without the added criterion of positive definiteness which is equivalent to assuming $(If,f) = 0 \Rightarrow f$ null. Bearing as it does a formal similarity to the assumption of completeness, this criterion will probably be no easier to verify in practice than completeness itself.

Problem

4.6 Show that there is a 'duality' between the two eigenvalue problems

(i) $\begin{cases} y'' + \lambda y = 0 \\ y(0) = y(1) = 0 \end{cases}$

and

(ii) $\lambda \int_0^1 T(x,t) f(t)\, dt = f(x),$

where T is the 'triangular' kernel

$$T(x,t) = \begin{cases} (1-x)t & (0 \leqslant t \leqslant x \leqslant 1) \\ x(1-t) & (0 \leqslant x \leqslant t \leqslant 1), \end{cases}$$

in the sense that they have the same spectrum and the same eigenfunctions.

For more on such duality properties see Lovitt (1950).

Appendix 1
Supplementary theorems

This appendix contains a list of various facts and principles which are needed in the text but which do not fit naturally into the presentation. Most of the details can be found in Yosida (1965).

1 The uniform boundedness principle *Let (T_n) be a sequence of bounded linear transformations of one Banach space B to another. For each $f \in B$, suppose there exists M_f such that $\|T_n f\| \leqslant M_f$ for every n. Then there exists M such that $\|T_n\| \leqslant M$.*

2 Properties of weakly convergent sequences *If (f_n) converges weakly to f in a Banach space, then*

(i) *there exists M such that $\|f_n\| < M$ for every n,*

and

(ii) *$f \in [f_n]$.*

3 The Weierstrass approximation theorems

(a) *Let f be a continuous complex valued function on the finite interval $[a, b]$. Then there exists a sequence of polynomials converging to f uniformly on $[a, b]$.*

(b) *Let f be a continuous complex valued function of period 2π defined on $\mathbb{R}$. Then there exists a sequence (p_n) of trigonometrical polynomials, i.e. $p_n(x) = \sum_{|k| \leqslant n} c_k e^{ikx}$, converging uniformly to f.*

4 The Riesz–Fréchet representation theorem *Let F be a bounded linear functional on a Hilbert space H. Then there exists $g_F \in H$ such that $Ff = (f, g_F)$ for every $f \in H$.*

Conversely, for fixed $g \in H$, (f, g) defines a bounded linear functional on H.

5 The 'bounded inverse' theorem *Let T be a bounded linear transformation, mapping a Banach space B onto another Banach space C in a one-to-one manner. Then T^{-1} exists as a bounded linear transformation of C onto B.*

6   Zorn's lemma

(*a*) Let r be a relation on a set S (i.e. a collection of ordered pairs of members of S; we write xry to mean $(x, y) \in r$) such that, for all x and y in S:

   (i) xrx;

   (ii) xry and $yrx \Rightarrow x = y$;

   (iii) xry and $yrz \Rightarrow xrz$.

Then S is said to be *partially ordered by r*.

If S is partially ordered by r and either xry or yrx for every $x, y \in S$, then S is said to be *linearly ordered*.

Let S be partially ordered by r. If $S' \subset S$, and there exists $z \in S$ such that xrz for every $x \in S'$, then z is called an *upper bound* for S'. If there exists $z \in S$ such that $zrx \Rightarrow x = z$, then z is called a *maximal element* of S.

(*b*) Zorn's lemma. *Let a non-empty set S be partially ordered by r. If every linearly ordered subset of S has an upper bound in S, then S has at least one maximal element.*

7   The Gram–Schmidt orthonormalisation process   Let

(u_n) be a sequence of linearly independent elements in a Hilbert space H. One forms an orthonormal sequence (v_n) by taking certain linear combinations of u_ns; there are two approaches:

(i) The 'geometrical' approach. We take $v_1 = \|u_1/u_1\|$ and define v_{n+1} inductively by subtracting from u_{n+1} its components along the previous (normalised) vs, viz.

$$v_{n+1} = \left(u_{n+1} - \sum_{k=1}^{n} (u_{n+1}, v_k)\, v_k \right) \Big/ \left\| \left(u_{n+1} - \sum_{k=1}^{n} (u_{n+1}, v_k)\, v_k \right) \right\|.$$

(ii) The 'linear algebra' approach. One expresses v_{n+1} as a linear combination of us, imposes the orthogonality conditions, and calculates the coefficients by Cramer's rule. The result is

$$v_{n+1} = \begin{vmatrix} (u_1, u_1) & \dots & (u_1, u_n) & u_1 \\ \vdots & & \vdots & \vdots \\ (u_{n+1}, u_1) & \dots & (u_{n+1}, u_n) & u_{n+1} \end{vmatrix} \bigg/ \sqrt{(G_n G_{n+1})}$$

where

$$G_n = \begin{vmatrix} (u_1, u_1) & \dots & (u_1, u_n) \\ \vdots & & \vdots \\ (u_n, u_1) & \dots & (u_n, u_n) \end{vmatrix}$$

and is called the *Gram determinant* of (u_n).

The same result is obtained from either approach, indeed *the ON sequence (v_n) is unique* subject to certain restrictions such as a fixed ordering of (u_n), etc.

8.   The extension of an operator 'by continuity' *Let H be a Hilbert space and let T be an operator defined on the dense set $D \subset H$. Syppose that $(f_n) \subset D$, and $f_n \to f \neq D$. Suppose also that $T f_n \to g$. Then we may extend T by the formula $T f = g$.*

9   The invertibility theorem of C. Neumann *Let T be a linear operator on a Hilbert space, and let I be the identity operator. Then if $\| I - T \| < 1$, T^{-1} exists, is bounded, and is unique.*

10   Useful theorems for the L^p theory of convergence and completeness   We have made no use of these in the text but they are sometimes useful and are included for reference. Theorem A is a theorem of F. Riesz; the two parts generalise the theorems of Parseval and Riesz–Fischer (q.v.) respectively. Theorem B is due to O. Szász (1947). In both theorems (a, b) is a finite interval of $\mathbb{R}$, and $p + q = pq$.

A *Let (ϕ_n) be an orthogonal sequence in $L^2(a, b)$ such that $|\phi_n| < M$ for every n. Let (a_n) be the sequence of Fourier coefficients of f with respect to (ϕ_n), and set $[\Sigma |a_n|^q]^{1/q} = \| a \|_q$. Let $1 < p \leqslant 2$. Then*

   (i) *if $f \in L^p(a, b)$, then*

$$\| a \|_q \leqslant M^{(2-p)/p} \| f \|_p ;$$

   (ii) *if $\| a \|_p < \infty$, then there exists $f \in L^q(a, b)$ such that*

$$\| f \|_q \leqslant M^{(2-p)/p} \| a \|_p .$$

See Zygmund (1959) p. 102 for necessary and sufficient conditions for equality to occur. Note that the results may fail if $p > 2$.

B. *Let $\phi(x)$ be bounded on $[a,b]$, and let $\phi(nx)$ $(n = 1, 2, ...)$ be orthogonal over (a,b) and complete on $L^p(a,b)$, $1 < p \leqslant 2$. Let $\psi \in L^q(a,b)$ and have Fourier coefficients (a_n) with respect to $(\phi(nx))$ such that (i) $a_m a_n = a_{mn}$ and (ii) $\Sigma |a_n|^p < \infty$. Then $(\psi(nx))$ is complete on $L^p(a,b)$.*

Appendix 2
Definitions of special functions

This list contains the definitions of most of the special functions used in the text; where there is a choice between several equivalent definitions that one which is most appropriate to the context has been preferred. For further information see, for example, Magnus *et al.* (1966).

1 Gamma function

$$\Gamma(z) = \int_0^\infty e^{-t} t^{z-1} dt, \quad \mathrm{Re}\, z > 0.$$

$\Gamma(z)\,\Gamma(1-z) = \pi/\sin \pi z$ (functional equation).

2 Pochhammer's symbol

$$(a)_n = a(a+1)\dots(a+n-1) = \frac{\Gamma(a+n)}{\Gamma(a)} \quad (n = 1, 2, \dots);$$

$$(a)_0 = 1.$$

3 Binomial coefficient

$$\binom{\alpha}{n} = \frac{\Gamma(1+\alpha)}{n!\,\Gamma(1+\alpha-n)} \quad (n = 1, 2, \dots)$$

4 Beta function

$$B(x,y) = (a-b)^{-x-y+1} \int_b^a (t-b)^{x-1}(a-t)^{y-1} dt$$
$$= \frac{\Gamma(x)\,\Gamma(y)}{\Gamma(x+y)}.$$

5 Jacobi polynomials

$$P_n^{(\alpha,\beta)}(x) = 2^{-n} \sum_{k=0}^n \binom{n+\alpha}{k}\binom{n+\beta}{n-k}$$
$$\times (x+1)^k (x-1)^{n-k} \quad (n = 0, 1, \dots).$$

Special cases are:

(i) $\alpha = \beta$, the Gegenbauer polynomials;

(ii) $\alpha = \beta = -\frac{1}{2}$, the Chebyshev polynomials of the 1st kind, $T_n(x)$;

(iii) $\alpha = \beta = \frac{1}{2}$, the Chebyshev polynomials of the 2nd kind, $U_n(x)$;

(iv) $\alpha = \beta = 0$, the Legendre polynomials $P_n(x)$.

6 Legendre functions

Legendre function: $P_z(x) = \dfrac{\sqrt{2}}{\pi} \displaystyle\int_0^{\cos^{-1}x} \dfrac{\cos(z+\frac{1}{2})t}{(\cos t - x)^{\frac{1}{2}}}\, dt.$

The special cases $z = 0, 1, \ldots$, yield the Legendre polynomials $P_n(x)$. Associated Legendre function:

$$P_n^m(x) = (1-x^2)^{\frac{1}{2}m} \frac{d^m}{dx^m} P_n(x),$$

m a positive integer.

7 Generalised Laguerre polynomials

$$L_n^{(\alpha)}(x) = \sum_{k=0}^{n} (-1)^k \binom{n+\alpha}{n-k} \frac{x^k}{k!}, \quad \alpha > -1.$$

The special cases $L_n^{(0)}(x) \equiv L_n(x)$ are called the Laguerre polynomials.

8 Hermite polynomials

$$H_n(x) = n! \sum_{k=0}^{\lceil n/2 \rceil} \frac{(-1)^k (2x)^{n-2k}}{k!(n-2k)!}.$$

9 Bessel function

$$J_\nu(x) = \frac{2(x/2)^\nu}{\Gamma(\nu+\frac{1}{2})\,\Gamma(\frac{1}{2})} \int_0^1 (1-t^2)^{\nu-\frac{1}{2}} \cos xt\, dt, \quad \nu > -\tfrac{1}{2}.$$

10 Struve function

$$\mathbf{H}_\nu(x) = \frac{2(x/2)^\nu}{\Gamma(\nu+\frac{1}{2})\,\Gamma(\frac{1}{2})} \int_0^1 (1-t^2)^{\nu-\frac{1}{2}} \sin xt\, dt, \quad \nu > -\tfrac{1}{2}.$$

11 Hypergeometric ${}_2F_1$

$$ {}_2F_1(a,b,c;x) = \sum_{k=0}^{\infty} \frac{(a)_k\,(b)_k}{(c)_k} \frac{x^k}{k!}.$$

Appendix 3
Some complete sequences of special functions

Most of the better known sets of special functions which form CON sequences in an L^2 space (or a subspace) are listed. L^p information can sometimes be obtained by using problem 2.2 or problem 1.8. It can also be found through the references to external sources; so can the definitions of the functions involved if they are not found in Appendix 2. Normalising factors are usually omitted; again, these can be traced through the references. The list is not intended to be exhaustive.

Properties of non-orthogonal sequences do not lend themselves easily to tabulation; references to more detailed information are given with each listing.

Orthogonal sequences			Complete in	Reference
Elementary functions	1	e^{inx}; $n = 0, \pm 1, \ldots$.	$L^2(-\pi, \pi)$	p. 36 and Martii (1969) p. 51
	2	$e^{2\pi inx}$; $n = 0, \pm 1, \ldots$.	$L^2(0, 1)$	p. 37
	3	$\sin\left[n\pi(x-a)/(b-a)\right]$; $n = 1, 2, \ldots$.	$L^2(a, b)$	p. 37
	4	$1, \cos\left[n\pi(x-a)/(b-a)\right]$; $n = 1, 2, \ldots$.	$L^2(a, b)$	p. 37
	5	$\cos^n x, \sin nx \cos^n x$; $n = 1, 2, \ldots$.	$L^2(0, 2\pi), \mu)$	p. 52
	6	$\sin\left(n-\tfrac{1}{2}\right)x$; $n = 1, 2, \ldots$.	$L^2(0, \pi)$	p. 106
	7	$\cos\left(n-\tfrac{1}{2}\right)x$; $n = 1, 2, \ldots$.	$L^2(0, \pi)$	p. 106
	8	$e^{-x}, \sin nx - n \cos nx$; $n = 1, 2, \ldots$.	$L^2(0, \pi)$	p. 106
Polynomials	9	$P_n^{(\alpha, \beta)}(x)$; $n = 0, 1, \ldots$.	$L^2((-1, 1), (1-x)^\alpha (1+x)^\beta)$	p. 33
	10	$P_n^m(x)$; $n = m, m+1, \ldots$.	$L^2(-1, 1)$	p. 38
	11	$L_n^{(\alpha)}(x)$; $n = 0, 1, \ldots$.	$L^2(\mathbb{R}^+, x e^{-x})$	p. 33
	12	$H_n(x)$; $n = 0, 1, \ldots$.	$L^2(\mathbb{R}, e^{-x^2})$	p. 33
	13	z^n; $n = 0, 1, \ldots, z = x + iy$.	$L^2_r(D)$	p. 65
	14	$U_n(z)$; $n = 0, 1, \ldots$.	$L^2_r(E)$	p. 68
	15	$p_n(r)$; $n = 0, 1, \ldots$.	$L^2(\mathbb{R}, \alpha)$	p. 43
Rational functions	16	$\dfrac{(ix-1)^{n+\rho}}{(ix+1)^{n+1+\rho}}$; $n = 0, \pm 1, \ldots$.	$L^2(\mathbb{R})$	p. 64
	17	$\left(\dfrac{a-ix}{a+ix}\right)^n$; $n = 0, \pm 1, \ldots$.	$L^2(\mathbb{R}, a/\pi(a^2+x^2))$	p. 64

[continued overleaf

Orthogonal sequences			Complete in	Reference
Higher transcendental functions	18	$x^{\frac{1}{2}}J_\nu(j_{n\nu}x)$; $n = 1, 2, \ldots$	$L^2(0, 1)$	p. 40
	19	$x^{\frac{1}{2}}I_\nu(\alpha_n x)$; $n = 1, 2, \ldots$, where $\alpha_n J_\nu'(\alpha_n) + hI_\nu(\alpha_n) = 0$, $h + \nu > 0$.	$L^2(0, 1)$	p. 42
	20	$x^{-\frac{1}{2}}J_{n+\frac{1}{2}}(x)$; $n = 0, 1, \ldots$	F^π	p. 59
	21	$x^{-\frac{1}{2}}J_{\nu+2n+1}(x)$; $n = 0, 1, \ldots$	B_ν	p. 59
	22	$se_n(x)$; $n = 1, 2, \ldots$	$L^2(0, \pi)$	p. 104
	23	$ce_n(x)$; $n = 0, 1, \ldots$	$L^2(0, \pi)$	p. 104
	24	$\{I_\nu(\lambda_n x)/I_\nu(\lambda_n) - J_\nu(\lambda_n x)/J_\nu(\lambda_n)\}$	$L^2(0, 1)$	Dahlberg (1973)
Discontinuous functions	25	$w_n(x)$; $n = 1, 2, \ldots$ (Walsh functions)	$L^2(0, 1)$	p. 47
	26	$h_n(x)$; $n = 1, 2, \ldots$ (Haar functions)	$L^2(0, 1)$	p. 49 and Singer (1970)
	27	'Square wave' functions.	$L^2(0, 2)$	Harrington and Cell (1961)
Interpolating functions	28	$\dfrac{\sin \pi(x-n)}{\pi(x-n)}$; $n = 0, \pm 1, \ldots$	F^π	p. 58
Higher dimensions	29	spherical harmonics	L^2 (sphere)	p. 38

Non-orthogonal sequences

			Information
Elementary functions	30	$\{e^{i\lambda_n x}\}$	§ 3.1.1.
	31	e^{inx}; $n = 1, 2, \ldots$	p. 87
	32	$1, \log|2 \sin n\pi x|$; $n = 1, 2, \ldots$	Complete on $L^p(0, \tfrac{1}{2})$, $p > 1$; Szász (1947)
Polynomials and powers	33	simple sets	p. 31
	34	z^n; $z = x+iy$, $n = 0, 1, \ldots$	p. 68
	35	x^n; $n = 0, 1, \ldots$	p. 29
	36	$\{x^{\lambda_n}\}$	§ 3.4.
Higher transcendental functions	37	$\{x^k J_\nu(\lambda_n x)\}$	p. 89
	38	$\{P_{\lambda_n}(x)\}$	p. 91
	39	$1, cn(mx, k)$; $m = 1, 2, \ldots$	$\left.\vphantom{\begin{array}{c}a\\b\end{array}}\right\}$ Complete in $L^2(0, 2K)$
	40	$sn(mx, k)$; $m = 1, 2, \ldots$ sn and cn are the Jacobian elliptic functions.	Craven (1971)
Piecewise linear functions	41	$\{\chi_r(x)\}$	p. 33
	42	$nx - [nx] - \tfrac{1}{2}$; $n = 1, 2, \ldots$ (Sawtooth functions)	Complete on $L^p(0, \tfrac{1}{2})$, $p > 1$; Szász (1947)
	43	$\mathrm{sgn}\,(\sin n\pi x)$; $n = 1, 2, \ldots$ ('square sine' waves)	Complete on $L^p(0, 1)$, $p > 1$; Szász (1947)
	44	Schauder system	Basis for $C[0, 1]$; Singer (1970), p. 11

Bibliography

Alexits, G. (1961). *Convergence problems of orthogonal series.* Pergamon, London.

Banach, S. (1932). *Théorie des opérations linéaires.* Warsaw.

Birkhoff, G. and Rota, G-C. (1969). *Ordinary differential equations.* Xerox College Publishing, Lexington.

Boas, R. P. and Pollard, H. (1947). Complete sets of Bessel and Legendre functions. *Annals of Math.* (2) **48**, 366-84.

Carlitz, L. (1960). Some orthogonal polynomials associated with elliptic functions. *Duke Math. J.* **27**, 443-59.

Coddington, E. A. and Levinson, N. (1955). *Theory of ordinary differential equations.* McGraw-Hill, New York.

Copson, E. T. (1967). *Metric spaces.* Cambridge University Press.

Craven, B. D. (1971). Stone's theorem and the completeness of orthogonal systems. *J. Austral. Math. Soc.* **12**, 211-23.

Dahlberg, E. (1973). On the completeness of eigenfunctions of some higher order operators. *J. Math. Anal. Appl.* **44**, 227-37.

Dalzell, D. P. (1945*a*). On the completeness of a series of normal orthogonal functions. *J. Lond. Math. Soc.* **20**, 87–93.

Dalzell, D. P. (1945*b*). On the completeness of Dini's series. *J. Lond. Math. Soc.* **20**, 213–18.

Duffin, R. J. and Eachus, J. J. (1942). Some notes on an expansion theorem of Paley and Wiener. *Bull. Amer. Math. Soc.* **48**, 850–5.

Enflo, P. (1973). A counterexample to the approximation problem in Banach spaces. *Acta Math.* **130**, 309-17.

Erdélyi, A. *et al.* (1953). *Higher transcendental functions,* vol. 3. McGraw-Hill, New York.

Graves, R. E. (1952). A closure criterion for orthogonal functions. *Canad. J. Math.* **4**, 198–203.

Hardy, G. H. (1941). Notes on special systems of orthogonal functions, I. *Proc. Cambridge Philos. Soc.* **37**, 331-48.

Hardy, G. H., Littlewood, J. E. and Polya, G. (1952). *Inequalities.* Cambridge University Press.

Harmuth, H. F. (1969). *Transmission of information by orthogonal functions.* Springer, Berlin.

Harrington, W. J. and Cell, J. W. (1961). A set of square wave functions orthogonal and complete in $L_2(0, 2)$. *Duke Math. J.* **28**, 393-407.

Higgins, J. R. (1972). An interpolation series associated with the Bessel–Hankel transform. *J. Lond. Math. Soc.* **5**, 707–14.

Higgins, J. R. (1976). A sampling theorem for irregularly spaced sample points. *IEEE Trans. Information Theory* IT-**22** (5), 621–3.

Higgins, J. R. *et al.* (1975). Problem 5960. *Amer. Math. Monthly* **82**, 859–60.

Hille, E. (1918). Some problems concerning spherical harmonics. *Arkiv för Matematik, Astronomi, och Fysik* **13**.

Kaczmarz, S. and Steinhaus, H. (1935). *Theorie der Orthogonalreihen.* Warsaw.

Kadec, M. I. (1964). The exact value of the Paley–Wiener constant. *Dokl. Akad. Nauk. SSSR* **155**, 1253–4 = *Soviet Math. Dokl.* **5**, 559–61.

Kanwal, R. P. (1971). *Linear integral equations.* Academic Press, New York.

Kato, T. (1966). *Perturbation theory for linear operators.* Springer, Berlin.

Lackey, R. B. and Meltzer, D. (1971). A simplified definition of Walsh functions. *IEEE Trans. Computers* C-**20**, 211–13.

Lehto, O. (1952). Some remarks on the kernel function in Hilbert function space. *Ann. Acad. Sci. Fenn. Ser. A*, no. 109.

Levinson, N. (1936*a*). On non-harmonic Fourier series. *Annals of Math.* **37**, 919–36.

Levinson, N. (1936*b*). On the closure of $\{e^{i\lambda_n x}\}$. *Duke Math. J.* **2**, 511–16.

Lovitt, W. V. (1950). *Linear integral equations.* Dover, New York.

Magnus, W., Oberhettinger, F. and Soni, R. P. (1966). *Formulas and theorems for the special functions of mathematical physics.* Springer, Berlin.

Marti, J. T. (1969). *Introduction to the theory of bases.* Springer, Berlin.

McLachlan, N. W. (1947). *Theory and applications of Mathieu functions.* Oxford University Press.

Meschkowski, H. (1962). *Hilbertsche Räume mit Kernfunktion.* Springer, Berlin.

Meux, J. W. (1966). Ordinary differential equations of the fourth order with orthogonal polynomial solutions. *Amer. Math. Monthly* **73**, 104–10.

Naimark, M. A. (1968). *Linear differential operators, part I.* Harrap, London.

Nehari, Z. (1952). *Conformal mapping.* McGraw-Hill, New York.

Nevai, G. P. (1973). Some properties of polynomials orthonormal with weight $(1+x^{2k})^{\alpha}e^{-x^{2k}}$, and their applications in approximation theory. *Dokl. Akad. Nauk. SSSR* **211** = *Soviet Math. Dokl.* **14**, 1116–19.

Olevskiĭ, A. M. (1975). *Fourier series with respect to general orthogonal series* (translated by B. P. Marshall and H. J. Christoffers). Ergebnisse der Mathematik und ihrer Grenzgebiete, Band 86. Springer, Berlin.

Paley, R. E. A. C. and Wiener, N. (1934). *Fourier transforms in the complex domain.* Amer. Math. Soc., New York.

Rainville, E. D. (1963). *Special functions.* MacMillan, New York.

Sagan, H. (1961). *Boundary and eigenvalue problems in mathematical physics.* Wiley, New York.

Sansone, G. (1959). *Orthogonal functions.* Interscience, New York.

Schmidt, E. (1933). Über die Charlier-Jordansche Entwicklung einer willkürlichen Funktion nach der Poissonschen Funktion und ihren Ableitungen. *Z. Angew. Math. Mech.* **13**, 139–42.

Simmons, G. F. (1963). *Topology and modern analysis.* McGraw-Hill, New York.

Singer, I. (1970). *Bases in Banach spaces.* Springer, Berlin.

Szász, O. (1947). On Möbius' inversion formula and closed sets of functions. *Trans. Amer. Math. Soc.* **62**, 213–39.

Szegö, G. (1939). *Orthogonal polynomials.* Amer. Math. S oc. New York.

Sz.-Nagy, B. (1965). *Introduction to real functions and orthogonal expansions.* Oxford University Press.

Titchmarsh, E. C. (1937). *Introduction to the theory of Fourier integrals.* Oxford University Press.

Titchmarsh, E. C. (1939). *The theory of functions.* Oxford University Press.

Titchmarsh, E. C. (1946). *Eigenfunction expansions associated with second order differential equations.* Oxford University Press.

Tricomi, F. G. (1955). Sulla chiusura dei sistemi ortonormali di funzioni. *Rev. Un. Mat. Argentina* **17**, 299–303.

Vogel, T. (1953). *Les fonctions orthogonales dans les problèmes aux limites de la Physique Mathematique.* Centre National de la Recherche Scientifique, Paris.

Watson, G. N. (1922). *Theory of Bessel functions.* Cambridge University Press.

Whittaker, E. T. (1915). On the functions which are represented by the expansions of the interpolation theory. *Proc. Roy. Soc. Edinburgh* **35**, 181–94.

Wiener, N. (1949). *Interpolation, extrapolation and smoothing of stationary time series.* The Technical Press of M.I.T. and Wiley, New York.

Yosida, K. (1965). *Functional analysis.* Spiinger, Berlin.

Young, R. M. (1975). A note on a trigonometric moment problem. *Proc. Amer. Math. Soc.* **49**, 411–15.

Zygmund, A. (1959). *Trigonometrical Series,* vol. 2. Cambridge University Press.

Index

associated Legendre equation, 110
associated Legendre functions, 38, 123
   definition of, 121

Banach space, 4
   dual, 8
   reflexive, 8, 9
   second dual, 8
basis
   absolute, 27
   Bessel, 74
   conditional, 27
   for Banach space, 26
   for Hilbert space, 20
   Hilbert, 74
   Riesz, 74
   unconditional, 27
   weak, 26
basis problem, 27
Bergman kernel, 66
Bessel function, 124, 125
   definition of, 121
Bessel–Neumann functions, 124
   completeness of, 59
Bessel's equation, 110
Bessel's inequality, 13
beta function, 91
   definition of, 120
binomial coefficient, definition of, 120
biorthogonal system, 19
biorthonormal (BON) system, 19
boundary value problem, 99
   regular self-adjoint, 108
   singular, 109
'bounded inverse' theorem, 23
   statement of, 117

cardinal series, 57
   interpolating functions of, 124

Cauchy sequence, 4
Chebyshev polynomials, first kind,
   definition of, 121
Chebyshev polynomials, second kind,
   68, 123
   definition of, 121
closed linear span, 3
complete orthonormal sequence,
   16
complete vector space, 4
completeness criterion
   Dalzell's, 39, 40
   Lauricella's, 34
   Vitali's, 35, 37, 38, 43, 69
completeness theorem
   for complex polynomials, 68
   for polynomials, 31
   of Rényi, 48
convergence, 3
   strong, 8
   weak, 8

Dalzell completeness criterion, 39
   modified form of, 40
dense subsets, chain of, 4
dimension (Hilbert), 18
direct sum, 6
discontinuous functions, 124

eigenfunctions, 98
   asymptotic form of, 103
   complete orthonormal property
   of, 103
eigenspace, 98
eigenvalue, 98
   asymptotic form of, 103
   problem, 100
eigenvector, 98
equivalent sequences, 74